MATHEMATICS ACTIVITIES
FOR ELEMENTARY TEACHERS

Seventh Edition

Dan Dolan
Project to Increase Mastery of Mathematics and Science,
Wesleyan University

Jim Williamson
The University of Montana

Mari Muri
Project to Increase Mastery of Mathematics and Science,
Wesleyan University

To accompany
Mathematical Reasoning for Elementary Teachers
Calvin Long/Duane DeTemple/Richard Millman

PEARSON

Boston Columbus Indianapolis New York San Francisco Upper Saddle River
Amsterdam Cape Town Dubai London Madrid Milan Munich Paris Montreal Toronto
Delhi Mexico City Sao Paulo Sydney Hong Kong Seoul Singapore Taipei Tokyo

Reproduced by Pearson Addison-Wesley from electronic files supplied by the author.

Copyright © 2015, 2013, 2010 Pearson Education, Inc.
Publishing as Pearson Addison-Wesley, 75 Arlington Street, Boston, MA 02116.

ISBN-13: 978-0-321-91511-5
ISBN-10: 0-321-91511-9

16 2023

Contents

Preface

ABOUT THIS BOOK   *Mathematics Activities for Elementary Teachers* provides a hands-on, manipulative-based, problem-solving approach to learning and teaching elementary mathematics. The activities in the book were developed to correspond to the content in *Mathematical Reasoning for Elementary Teachers*, Seventh Edition by Calvin T. Long, Duane W. DeTemple, and Richard S. Millman (Pearson Education, Inc. 2015). Although this activities manual was designed to supplement the textbook, it can be used to develop students' understanding of mathematical concepts and skills in a variety of settings:

- mathematics content courses for preservice elementary teachers, grades K-8,
- mathematics methods courses for preservice elementary teachers, grades K-8, and
- inservice courses and professional development workshops for elementary and middle school teachers.

Students' involvement in these activities demonstrates an alternative approach to the traditional teaching and learning of mathematics. The activities are based on constructing understanding through interactive experiences and sequenced in a developmentally appropriate manner. The activities can be used to:

- develop mathematical concepts or procedures;
- reinforce concepts that have been previously taught;
- illustrate applications of mathematical concepts in contextual situations; and
- promote construction of students' knowledge of mathematics.

While this book has been designed primarily for preservice and inservice classes, it also serves as an excellent source of activities for elementary teachers at various grade levels to include in their mathematics instruction in grades K through 8.

A RESEARCH-BASED APPROACH   In the *Effective Instruction Clips* (NCTM, 2010) it states that

> "Two features of instruction are especially likely to help students develop conceptual understanding of the mathematics topics they are studying:
> - attending explicitly to connections among facts, procedures, and ideas, and
> - encouraging students to wrestle with the important ideas in an intentional and conscious way.

In essence, if instruction aims to help students develop conceptual understanding, then it must make explicit the crucial relationships that lie at the heart of such understanding.

Research findings suggest the following: mathematics teaching that facilitates skill efficiency

- is rapidly paced;
- includes modeling by the teacher with many teacher-directed product type of questions;
- displays a smooth transition from demonstration to substantial amounts of error-free practice.

The teacher plays a central role in organizing, pacing, and presenting information to meet well-defined learning goals."

"The kinds of experiences teachers provide clearly play a major role in determining the extent and quality of students' learning."

Principles and Standards for School Mathematics
NCTM, 2000

The *Principles and Standards for School Mathematics* (National Council of Teachers of Mathematics, 2000) describes knowing mathematics as having the ability to use it in meaningful ways. In the process of learning mathematics, teachers must be involved in doing mathematics–investigating, conjecturing, discussing, and validating–in order to develop confidence in their own mathematical ability and to be able to instill an appreciation of its value in their students. "They (teachers) need to know the ideas with which students often have difficulty and ways to help bridge common misunderstandings."

These mathematics reform documents call for a basic restructuring of the curriculum, instructional practices, and assessment systems for mathematics in grades K-16, and of the program for the preparation of teachers of mathematics. This transformation, which is echoed in the *Standards for Mathematical Practice* of the *Common Core State Standards for Mathematics* (National Governors Association Center for Best Practices and Council of Chief State School Officers, 2010), necessitates dramatic changes in the ways that prospective teachers learn and later teach mathematics. As noted in the document, *A Call for Change; Recommendations for the Preparation of Elementary Teachers* (Mathematical Association of America, 1991), alternative methods must be presented to preservice and inservice teachers so that they can learn and practice them while they are in the learning process themselves.

GOALS OF THE BOOK

This book was written to engage preservice and inservice teachers in doing mathematics rather than simply reading about mathematics. It is not intended to be a textbook for a mathematics content course at the collegiate level or to provide all of the content necessary for such a course. Rather, it is intended to be used as a companion to a textbook, such as *Mathematical Reasoning for Elementary Teachers,* to provide hands-on, problem-based activities and attention to the *Standards for Mathematical Practice* that involve preservice elementary teachers in discovering mathematical concepts, doing real problem solving, and exploring mathematical concepts in interesting, stimulating, and real-world settings. All of these activities can then be adapted for use with elementary students when the preservice teachers enter the profession.

The content and instructional approach of this activities manual embodies the spirit and intent of the NCTM Standards documents, as well as the *Common Core State Standards for Mathematics* and the *Standards for Mathematical Practice*. In the process of engaging students in meaningful mathematical tasks, group work, hands-on activities, and classroom discourse, these materials capture the essence of the first standard for the professional development of teachers in the *Professional Teaching Standards for School Mathematics*, (NCTM, 1991).

"Mathematics and mathematics education instructors in preservice and continuing education programs should model good mathematics teaching by

- posing worthwhile mathematical tasks;
- engaging teachers in mathematical discourse;
- enhancing mathematical discourse through the use of a variety of tools, including calculators, computers, and physical and pictorial models;
- creating learning environments that support and encourage mathematical reasoning and teachers' dispositions and abilities to do mathematics;
- expecting and encouraging teachers to take intellectual risks in doing mathematics and to work independently and collaboratively;
- representing mathematics as an ongoing human activity; and
- affirming and supporting full participation and continued study of mathematics by all students."

If elementary teachers learn concepts through a problem-solving activity approach, develop ideas from the concrete level to the abstract level, and connect multiple representations of mathematical ideas, they will gain a broader and deeper understanding of mathematics and begin to construct their own meanings of mathematical concepts. As a result, they will be better prepared to create rich mathematical learning environments in their classrooms for students.

The activities in this book promote the major goals of the NCTM Standards documents. These goals state that students should:

- become mathematical problem solvers,
- learn to communicate mathematically,
- learn to reason mathematically,
- become confident in their ability to do mathematics, and
- learn to value mathematics.

These activities also support the eight *Standards for Mathematical Practice* as described in the *Common Core State Standards for Mathematics*.

1. Make sense of problems and persevere in solving them.
2. Reason abstractly and quantitatively.
3. Construct viable arguments and critique the reasoning of others.
4. Model with mathematics.
5. Use appropriate tools strategically.
6. Attend to precision.
7. Look for and make use of structure.
8. Look for and express regularity in repeated reasoning.

These goals are consciously reflected in every activity in this book. All of the activities have been tested extensively with preservice teachers, teachers in professional development programs, and with K–8 students. They have proven effective and stimulating with all groups. We hope that as you engage in the activities in your preservice or inservice program, and when you use them with your students, you will enjoy exploring mathematics, become confident in doing mathematics, and will come to love mathematics as we do.

Introduction

CONTENT

"As students attempt to solve rich problem tasks, they come to understand the mathematical concepts and methods, become more adept at mathematical problem solving, and develop mathematical habits of mind that are useful ways to think about any mathematical situation."

Teaching Mathematics through Problem Solving
NCTM, 2003

This book contains activities designed to provide preservice teachers, inservice teachers, and teachers engaged in professional development programs opportunities to explore mathematical ideas and to develop conceptual understanding using a problem-solving approach and a variety of manipulative materials and instructional formats.

A problem-solving approach to learning enhances the development, reinforcement, and application of mathematical concepts in a way that practicing rote skills cannot. This approach also engages preservice teachers in learning mathematics in a way we trust they will apply when teaching their students.

For the most part, the activities do not require expensive or single-use materials. Full-color replicas of some commercial materials that can be used at many levels and in a variety of settings are included in the back of the book.

The *Instructor's Resource Guide,* available online, contains answers for all activities as well as some additional teaching suggestions.

ORGANIZATION

"Students learn mathematics through the experiences that teachers provide. Thus, students' understanding of mathematics, their ability to use it to solve problems, and their confidence in, and disposition toward mathematics are all shaped by the teaching they encounter in school. The improvement of mathematics education for all students requires effective mathematics teaching in all classrooms."

Principles and Standards for School Mathematics
NCTM, 2000

The instructional plan that precedes each activity provides direction for the course instructor and includes the following elements:

- **Purpose** outlines the mathematical concepts that are developed in the activity and describes which learning objectives—introducing, developing, reinforcing, or applying—the activity is designed to meet.

- **Common Core SMP** lists the *Standards for Mathematical Practice* (SMPs) that are incorporated in each activity. A complete listing and description of the SMPs is included in the *Common Core State Standards* section of this book.

- **Materials** describes any special equipment or supplies that are needed for the activity. Models of several manipulatives are in the Pouch in the back of the book. Activity Masters for use with various activities are available online.

- **Grouping** describes the classroom setting for the activity. Many activities can be completed on an individual basis. However, we encourage working in pairs or small groups whenever appropriate in order to model students working collaboratively as recommended in *A Call for Change* and the *Principles and Standards for School Mathematics,* and to model how mathematics should be taught in the elementary grades. See the section *Collaborative Learning* following the Introduction.

- **Getting Started** provides an introduction to the activity or explains any preparation necessary before using it. Since the purpose of these activities is not just to teach mathematical concepts, but also to illustrate how the concepts should be taught in the elementary grades, it is important that preservice and inservice teachers keep the following questions in mind and formulate answers to them as they are completing each activity:

  A. How could this activity be used with elementary students?
  B. At which grade level would the activity be appropriate?
  C. What adaptations would be necessary to make the activity suitable for use with elementary students at various grade levels?
  D. How could this activity be modified to differentiate instruction in a mixed ability classroom?
  E. How is this activity important to the development of students' mathematical understanding?
  F. How can I assure the incorporation of the SMPs into my lesson?
  G. What assessment methods could be used to determine students' understanding?

- **Extensions** present suggestions and ideas for extending the activity to other mathematical topics or making connections between the mathematical concepts in the activity and their application in the real world or other curricular areas. Included are some or all of the following:

  A. additional questions and problems to explore,
  B. questions to extend ideas in selected problems or the entire activity, and
  C. other activities, problems, and information related to the concepts in the activity.

Once the activity has been completed, the instructor should review and extend the activity by facilitating whole class discussion of the results as called for in SMP 3 and SMP 6, summarizing the activity, formalizing the mathematical content, and discussing its use with elementary students.

TIME REQUIREMENTS   The activities in this book can replace much of the lecture time that is usually devoted to teaching topics. Many concepts and skills are developed in the activities by working from the concrete level, to the representational level (connecting models and symbolism per SMP 4), to the abstract level. Through a carefully guided discussion of the results of an activity, students will develop their own mental constructs of the concepts being presented.

MATERIALS   **Manipulatives:** (In the **Pouch** in the back of the book) These are printed one-sided on heavy stock and in color to match the commercial products. You can punch them out for use with many activities. We suggest that you store each individual set in a plastic zip-lock bag or envelope.

Activity Masters (Available **Online** at www.pearsonhighered.com/dolan.) These consist of models, activity recording sheets, graph paper, nets for polyhedrons, activity cards, etc. that are to be used as part of selected activities. They should be printed for use with the activity.

Other Some activities require the use of typical mathematical tools – ruler, centimeter tape, calculator, computer, a protractor, etc. Others will use such items as scissors, dice, a deck of cards, etc., per SMP 5. The instructor may provide these or you will need to obtain them before you begin the activity.

Common Core State Standards for Mathematics

The *Common Core State Standards for Mathematics* (CCSSM) were developed through a state spearheaded initiative. Their intent is to define the mathematics necessary for students' career and college readiness. The CCSSM have been adopted as the statewide set of curriculum standards by almost every state in the nation. The CCSSM are comprised of two major sections:

1. Standards for Mathematical Content
2. Standards for Mathematical Practice

The textbook, *Mathematical Reasoning for Elementary Teachers, 7th Edition,* has been aligned to the *Standards for Mathematical Content* and each activity in *Mathematics Activities for Elementary Teachers* has been coded to the *Standards for Mathematical Practice* (SMPs) that will be used to complete the problems and exercises.

The SMPs are loosely based on the NCTM process standards of communication, connections, problem solving, reasoning and proof, and representation. The process standards have been expanded and improved upon to create the current set of SMPs. The SMPs describe ways in which students can more deeply engage with the mathematics they will encounter in the elementary and middle school grades. It is incumbent upon teachers to meld the mathematics content with the SMPs.

The eight SMPs are shown in their entirety at the end of this section. You will also want to read the Collaborative Learning section which will provide suggestions on how to have students engage in some of these SMPs naturally. The classroom of today should no longer see a teacher lecturing and doing sample problems on the board with a follow-up of students engaged in practicing the same exact types of problems. Rather, students need to engage in rich classroom discussion with a partner, in small groups, and with the entire class. They need to listen to each other, make sense of what they hear, and engage in dialogue about the merits of mathematical ideas or assumptions brought forward. The teacher can no longer be solely the "Sage on the Stage," but rather, the "Guide on the Side".

States that have adopted the CCSSM have placed an emphasis on the SMPs. Only if the classroom environment changes to incorporate the practice standards, can we expect students to assimilate and do well with the *Standards for Mathematical Content*. It is likely not uncommon for the practice standards to spill over into other content areas. Students listening to each other and critiquing each others' work, working on a problem long enough to see it to completion, using reasoning and appropriate tools where possible are all examples of how these standards can improve teaching and learning across the curriculum.

COMMON CORE STATE STANDARDS FOR MATHEMATICS STANDARDS FOR MATHEMATICAL PRACTICE

1. Make sense of problems and persevere in solving them.

Mathematically proficient students start by explaining to themselves the meaning of a problem and looking for entry points to its solution. They analyze givens, constraints, relationships, and goals. They make conjectures about the form and meaning of the solution and plan a solution pathway rather than simply jumping into a solution attempt. They consider analogous problems, and try special cases and simpler forms of the original problem in order to gain insight into its solution. They monitor and evaluate their progress and change course if necessary. Older students might, depending on the context of the problem, transform algebraic expressions or change the viewing window on their graphing calculator to get the information they need. Mathematically proficient students can explain correspondences between equations, verbal descriptions, tables, and graphs or draw diagrams of important features and relationships, graph data, and search for regularity or trends. Younger students might rely on using concrete objects or pictures to help conceptualize and solve a problem. Mathematically proficient students check their answers to problems using a different method, and they continually ask themselves, "Does this make sense?" They can understand the approaches of others to solving complex problems and identify correspondences between different approaches.

2. Reason abstractly and quantitatively.

Mathematically proficient students make sense of quantities and their relationships in problem situations. They bring two complementary abilities to bear on problems involving quantitative relationships: the ability to *decontextualize*—to abstract a given situation and represent it symbolically and manipulate the representing symbols as if they have a life of their own, without necessarily attending to their referents—and the ability to *contextualize*, to pause as needed during the manipulation process in order to probe into the referents for the symbols involved. Quantitative reasoning entails habits of creating a coherent representation of the problem at hand; considering the units involved; attending to the meaning of quantities, not just how to compute them; and knowing and flexibly using different properties of operations and objects.

3. Construct viable arguments and critique the reasoning of others.

Mathematically proficient students understand and use stated assumptions, definitions, and previously established results in constructing arguments. They make conjectures and build a logical progression of statements to explore the truth of their conjectures. They are able to analyze situations by breaking them into cases, and can recognize and use counterexamples. They justify their conclusions, communicate them to others, and respond to the arguments of others. They reason inductively about data, making plausible arguments that take into account the context from which the data arose. Mathematically proficient students are also able to compare the effectiveness of two plausible arguments, distinguish correct logic or reasoning from that which is flawed, and—if there is a flaw in an argument—explain what it is. Elementary students can construct arguments using concrete referents such as objects, drawings, diagrams, and actions. Such arguments can make sense and be correct, even though they are not generalized or made formal until later grades. Later, students learn to determine domains to which an argument applies. Students at all grades can listen or read the arguments of others, decide whether they make sense, and ask useful questions to clarify or improve the arguments.

4. Model with mathematics.

Mathematically proficient students can apply the mathematics they know to solve problems arising in everyday life, society, and the workplace. In early grades, this might be as simple as writing an addition equation to describe a situation. In middle grades, a student might apply proportional reasoning to plan a school event or analyze a problem in the community. By high school, a student might use geometry to solve a design problem or use a function to describe how one quantity of interest depends on another. Mathematically proficient students who can apply what they know are comfortable making assumptions and approximations to simplify a complicated situation, realizing that these may need revision later. They are able to identify important quantities in a practical situation and map their relationships using such tools as diagrams, two-way tables, graphs, flowcharts and formulas. They can analyze those relationships mathematically to draw conclusions. They routinely interpret their mathematical results in the context of the situation and reflect on whether the results make sense, possibly improving the model if it has not served its purpose.

5. Use appropriate tools strategically.

Mathematically proficient students consider the available tools when solving a mathematical problem. These tools might include pencil and paper, concrete models, a ruler, a protractor, a calculator, a spreadsheet, a computer algebra system, a statistical package, or dynamic geometry software. Proficient students are sufficiently familiar with tools appropriate for their grade or course to make sound decisions about when each of these tools might be helpful, recognizing both the insight to be gained and their limitations. For example, mathematically proficient high school students analyze graphs of functions and solutions generated using a graphing calculator. They detect possible errors by strategically using estimation and other mathematical knowledge. When making mathematical models, they know that technology can enable them to visualize the results of varying assumptions, explore consequences, and compare predictions with data. Mathematically proficient students at various grade levels are able to identify relevant external mathematical resources, such as digital content located on a website, and use them to pose or solve problems. They are able to use technological tools to explore and deepen their understanding of concepts.

6. Attend to precision.

Mathematically proficient students try to communicate precisely to others. They try to use clear definitions in discussion with others and in their own reasoning. They state the meaning of the symbols they choose, including using the equal sign consistently and appropriately. They are careful about specifying units of measure, and labeling axes to clarify the correspondence with quantities in a problem. They calculate accurately and efficiently, express numerical answers with a degree of precision appropriate for the problem context. In the elementary grades, students give carefully formulated explanations to each other. By the time they reach high school they have learned to examine claims and make explicit use of definitions.

7. Look for and make use of structure.

Mathematically proficient students look closely to discern a pattern or structure. Young students, for example, might notice that three and seven more is the same amount as seven and three more, or they may sort a collection of shapes according to how many sides the shapes have. Later, students will see 7×8 equals the well remembered $7 \times 5 + 7 \times 3$, in preparation for learning about the distributive property. In the expression $x^2 + 9x + 14$, older students can see the 14 as 2×7 and the 9 as $2 + 7$. They recognize the significance of an existing line in a geometric figure and can use the strategy of drawing an auxiliary line for solving problems. They also can step back for an overview and shift perspective. They can see complicated things, such as some algebraic expressions, as single objects or as being composed of several objects. For example, they can see $5 - 3(x - y)^2$ as 5 minus a positive number times a square and use that to realize that its value cannot be more than 5 for any real numbers x and y.

8. Look for and express regularity in repeated reasoning.

Mathematically proficient students notice if calculations are repeated, and look both for general methods and for shortcuts. Upper elementary students might notice when dividing 25 by 11 that they are repeating the same calculations over and over again, and conclude they have a repeating decimal. By paying attention to the calculation of slope as they repeatedly check whether points are on the line through $(1, 2)$ with slope 3, middle school students might abstract the equation $(y - 2)/(x - 1) = 3$. Noticing the regularity in the way terms cancel when expanding $(x - 1)(x + 1)$, $(x - 1)(x^2 + x + 1)$, and $(x - 1)(x^3 + x2 + x + 1)$ might lead them to the general formula for the sum of a geometric series. As they work to solve a problem, mathematically proficient students maintain oversight of the process, while attending to the details. They continually evaluate the reasonableness of their intermediate results.

Source: www.corestandards.org/Math/Practice

© Copyright 2010. National Governors Association Center for Best Practices and Council of Chief State School Officers. All rights reserved.

Correlation of *Mathematics Activities for Elementary Teachers* to the Common Core Standards of Mathematical Practice

Chapter	Activity Number and Title	Standards of Mathematical Practice
1	1: When You Don't Know What to Do	SMP 1, SMP 4, SMP 5, SMP 7, SMP 8
	2: What's the Pattern?	SMP 4, SMP 7, SMP 8
	3: Fascinating Fibonacci	SMP 1, SMP 6, SMP 7
	4: What's the Rule?	SMP 4, SMP 7, SMP 8
	5: Ten People in a Canoe	SMP 1, SMP 4, SMP 7
	6: Magic Number Tricks	SMP 1, SMP 2, SMP 8
	7: An Ancient Game	SMP 1, SMP 8
	8: What's the Number?	SMP 1, SMP 4, SMP 7
	9: Eliminate the Impossible	SMP 1, SMP 7, SMP 8
2	1: What's in the Loop?	SMP 1, SMP 2, SMP 3, SMP 4, SMP 6, SMP 7
	2: Loop de Loops	SMP 1, SMP 2, SMP 3, SMP 5, SMP 6, SMP 7
	3: Odd and Even Patterns	SMP 4, SMP 5, SMP 7, SMP 8
	4: Multiplication Arrays	SMP 6, SMP 7, SMP 8
	5: How Many Cookies?	SMP 6, SMP 7, SMP 8
	6: Find the Missing Factor	SMP 4, SMP 6, SMP 7
	7: Paper Powers	SMP 4, SMP 6, SMP 7
	8: The King's Problem	SMP 1, SMP 4, SMP 5, SMP 6, SMP 7
3	1: Regrouping Numbers	SMP 3, SMP 4, SMP 5, SMP 7
	2: Find the Missing Numbers	SMP 1, SMP 2, SMP 4, SMP 5, SMP 6
	3: It All Adds Up	SMP 4, SMP 5, SMP 6, SMP 7
	4: What's the Difference?	SMP 4, SMP 5, SMP 6, SMP 7
	5: Multi-digit Multiplication	SMP 4, SMP 5, SMP 6, SMP 7
	6: Least and Greatest	SMP 1, SMP 3, SMP 6, SMP 7, SMP 8
	7: Multi-digit Division	SMP 4, SMP 5, SMP 6, SMP 7
	8: Target Number	SMP 1, SMP 4, SMP 5
	9: A Visit to Fouria	SMP 1, SMP 2, SMP 4, SMP 5, SMP 6
4	1: A Square Experiment	SMP 3, SMP 4, SMP 5, SMP 7
	2: The Factor Game	SMP 2, SMP 4, SMP 5, SMP 7, SMP 8
	3: A Sieve of Another Sort	SMP 2, SMP 3, SMP 4, SMP 5, SMP 7, SMP 8
	4: Great Divide Game	SMP 5, SMP 6, SMP 7
	5: Tiling with Squares	SMP 2, SMP 3, SMP 4
	6: Pool Factors	SMP 4, SMP 5, SMP 6, SMP 7
5	1: Charged Particles	SMP 4
	2: Coin Counters	SMP 2, SMP 3, SMP 4, SMP 5
	3: Subtraction Patterns	SMP 1, SMP 2, SMP 7
	4: A Clown on a Tightrope	SMP 2, SMP 4, SMP 5
	5: Multiplication and Division Patterns	SMP 1, SMP 2, SMP 7
	6: Integer $\times$ and $\div$ Contig	SMP 2, SMP 7

Correlation of *Mathematics Activities for Elementary Teachers* to the Common Core Standards of Mathematical Practice

Chapter	Activity Number and Title	Standards of Mathematical Practice
6	1: What is a Fraction?	SMP 4, SMP 5
	2: Square Fractions	SMP 1, SMP 4, SMP 5
	3: Equivalent Fractions	SMP 4, SMP 5
	4: How Big Is It?	SMP 2, SMP 3, SMP 6
	5: What Comes First?	SMP 2, SMP 7
	6: Fraction War	SMP 2, SMP 3
	7: Adding and Subtracting Fractions	SMP 2, SMP 5, SMP 6
	8: Multiplying Fractions	SMP 2, SMP 5, SMP 6
	9: Dividing Fractions	SMP 2, SMP 5, SMP 6
7	1: What's My Name?	SMP 2, SMP 5
	2: Repeating Decimals	SMP 1, SMP 2, SMP 5, SMP 8
	3: Race for the Flat	SMP 4, SMP 5
	4: Empty the Board	SMP 1, SMP 4, SMP 5
	5: Deci-Order	SMP 4, SMP 5, SMP 7
	6: Decimal Arrays	SMP 3, SMP 4, SMP 5
	7: Decimal Multiplication	SMP 3, SMP 4, SMP 5
	8: Dice and Decimals	SMP 3, SMP 4, SMP 5
	9: Professors Short and Tall	SMP 3, SMP 5, SMP 6
	10: What Is Percent?	SMP 1, SMP 3, SMP 4, SMP 5
8	1: Patterns and Expressions	SMP 1, SMP 4, SMP 5, SMP 7, SMP 8
	2: Regular Polygons in a Row	SMP 3, SMP 7, SMP 8
	3: What's My Function?	SMP 1, SMP 2, SMP 4, SMP 5
	4: Exploring Linear Functions	SMP 1, SMP 2, SMP 3, SMP 4, SMP 5, SMP 8
	5: Graphing Rectangles	SMP 1, SMP 2, SMP 3, SMP 4, SMP 5, SMP 6, SMP 7, SMP 8
9	1: What's the Angle?	SMP 3, SMP 4, SMP 5, SMP 6, SMP 7, SMP 8
	2: Triangle Properties - Angles	SMP 3, SMP 4, SMP 5, SMP 6, SMP 7, SMP 8
	3: Inside or Outside?	SMP 2, SMP 4, SMP 5, SMP 6, SMP 7, SMP 8
	4: Angles on Pattern Blocks	SMP 2, SMP 4, SMP 5, SMP 6, SMP 7, SMP 8
	5: Sum of Interior/Exterior Angles	SMP 2, SMP 4, SMP 5, SMP 6, SMP 7
	6: Stars and Angles	SMP 2, SMP 4, SMP 5, SMP 6, SMP 7
	7: Mysterious Midpoints	SMP 2, SMP 3, SMP 4, SMP 5, SMP 6, SMP 7, SMP 8
	8: Spatial Visualization	SMP 2, SMP 3, SMP 4, SMP 5, SMP 6, SMP 7, SMP 8

Correlation of *Mathematics Activities for Elementary Teachers* to the Common Core Standards of Mathematical Practice

Chapter	Activity Number and Title	Standards of Mathematical Practice
10	1: Areas of Polygons	SMP 3, SMP 4, SMP 5, SMP 7
	2: From Rectangles to Parallelograms	SMP 3, SMP 4, SMP 5, SMP 7
	3: From Parallelograms to Triangles	SMP 3, SMP 4, SMP 5, SMP 7
	4: From Parallelograms to Trapezoids	SMP 3, SMP 4, SMP 5, SMP 7
	5: Pythagorean Puzzles	SMP 3, SMP 4, SMP 5, SMP 7
	6: Now You See It, Now You Don't	SMP 2, SMP 3, SMP 4, SMP 5, SMP 7
	7: Right or Not?	SMP 2, SMP 3, SMP 4, SMP 5, SMP 7
	8: Volume of a Rectangular Solid	SMP 2, SMP 3, SMP 4, SMP 5, SMP 6, SMP 7
	9: Pyramids and Cones	SMP 3, SMP 4, SMP 5, SMP 6, SMP 7
	10: Surface Area	SMP 3, SMP 4, SMP 5, SMP 6, SMP 7
11	1: Reflections	SMP 2, SMP 3, SMP 4, SMP 5, SMP 6, SMP 7
	2: Glide Reflections	SMP 4, SMP 5, SMP 7
	3: Translations	SMP 2, SMP 4, SMP 5, SMP 7
	4: Rotations	SMP 2, SMP 3, SMP 5, SMP 7
	5: Dilations	SMP 3, SMP 4, SMP 5, SMP 7
	6: Draw It	SMP 4, SMP 5, SMP 7
	7: Tessellations	SMP 2, SMP 3, SMP 4, SMP 5, SMP 6, SMP 7
12	1: Triangle Properties - Sides	SMP 2, SMP 3, SMP 4, SMP 5, SMP 6, SMP 7
	2: To Be or Not to Be Congruent	SMP 2, SMP 3, SMP 4, SMP 5, SMP 6, SMP 7
	3: Pattern Block Similarity	SMP 2, SMP 3, SMP 4, SMP 5, SMP 6, SMP 7
	4: Similar Triangles	SMP 3, SMP 4, SMP 5, SMP 7
	5: Outdoor Geometry	SMP 2, SMP 3, SMP 4, SMP 5, SMP 6, SMP 7, SMP 8
	6: Side Splitter Theorem	SMP 3, SMP 4, SMP 5, SMP 7, SMP 8
13	1: Graphing m&m's	SMP 3, SMP 4, SMP 5, SMP 7
	2: Grouped Data	SMP 2, SMP 3, SMP 4, SMP 7
	3: What's the Average?	SMP 2, SMP 3, SMP 4, SMP 5, SMP 7
	4: Finger Snapping Time	SMP 2, SMP 3, SMP 4, SMP 5, SMP 7
	5: The Weather Report	SMP 1, SMP 2, SMP 3, SMP 4, SMP 5, SMP 7
	6: Populations and Samples	SMP 1, SMP 2, SMP 3, SMP 4, SMP 5, SMP 7
14	1: What Are the Chances?	SMP 2, SMP 3, SMP 4, SMP 5
	2: The Spinner game	SMP 2, SMP 3, SMP 4, SMP 5, SMP 7
	3: Theoretical Probability	SMP 2, SMP 3, SMP 4, SMP 5, SMP 6, SMP 7, SMP 8
	4: Paper-Scissor-Rock	SMP 2, SMP 3, SMP 4, SMP 5, SMP 7, SMP 8
	5: How Many Arrangements?	SMP 1, SMP 2, SMP 3, SMP 4, SMP 5, SMP 7, SMP 8
	6: Pascal's Probabilities	SMP 2, SMP 3, SMP 4, SMP 5, SMP 7
	7: Simulate It	SMP1, SMP 2, SMP 3, SMP 4, SMP 5, SMP 7, SMP 8

Collaborative Learning

WHAT IS IT?   Collaborative learning, also described as *cooperative learning* or *cooperative grouping*, is any setting where individuals come together to share their talents to solve a problem. Teachers select groups of students to work together based on a variety of factors.

A teacher may select students to work together in order to differentiate instruction. Students of like cognitive ability work together to solve problems that were selected based on each group's ability level. Each group works on a problem that will challenge its ability. However, this grouping strategy does not take into account the various learning styles or mathematical strengths of students.

Arranging students within a group based on different mathematical strengths or different learning styles is also an effective method of getting students to work collaboratively. These grouping strategies allow students to share their diverse ways of thinking. This is beneficial to students of all ability levels. Brighter students become aware of alternative ways of looking at and solving problems. Lesser ability students are exposed to more sophisticated problem solving methods of their peers, and average students learn from both kinds of students. All students, regardless of ability level or learning style, have the opportunity to assimilate a variety of strategies into their problem-solving repertoire.

No one grouping strategy should be used exclusively. All grouping strategies require that teachers must know their students well in order to make the best selection for the assigned task.

WHY USE IT?   With the inception of the *Common Core State Standards*, which include the *Standards for Mathematical Practice*, it has become even more important to have students working collaboratively to solve problems. SMP 3 requires that students engage in "constructing viable arguments and critiquing the reasoning of others." The practice for this kind of communication takes place best in small group settings before sharing with the entire class. Another benefit of working collaboratively is to "attend to precision" as in SMP 6 where students are urged to use the precise mathematical language in presenting and/or defending their views or solutions.

Collaborative learning can improve students' academic achievement, help them build a greater sense of self-confidence, and enhance their social skills. Perhaps the most compelling reason for engaging students in collaborative problem solving is that it will prepare them for the world of work.

Successful corporations rely on a team approach. Manufacturing teams developing new products work collaboratively and each team member has an assigned role based on his or her background and specific skills. The team meets to discuss the problems they face. Group discussions promote diverse thinking and a variety of problem solving strategies. In the end, the product produced belongs to all who contributed to its development. Thus, each team member shares in the pride of the completed product.

WHAT DOES IT LOOK LIKE?

Researchers have analyzed and described a variety of grouping strategies to promote collaborative learning. No one method will work in all cases, but the interdependence of group members is a critical factor in the success of collaborative learning. Teachers need to select group members based on their specific skills and strengths and to assure that the group has enough talent among its members to accomplish the task. The group's responsibility is to take advantage of what each person contributes to the solution of the problem.

Group size can vary. Students can work effectively in pairs. The pair may be of different ability levels (one brighter and one slower student) or have different mathematical strengths (one student with strong spatial sense and the other with good numerical skills). Students may also have different learning styles (one kinesthetic learner and the other auditory). These pairs of students can pool their talents to solve a problem. Paired work is the basis for a cooperative strategy known as "think, pair, share." The teacher poses a problem; students are asked to think individually about the problem and then share this thinking with a classmate. One student from each pair then reports to the class.

Group size can also range from three to five students. When group size is larger than two, specific roles are assigned to ensure that each student has an opportunity to participate. Roles may include recorder, questioner, materials manager, reporter, etc. Upon completion of the task, each group member is responsible for knowing the solution to the problem.

CHALLENGES FOR THE TEACHER

To facilitate the collaborative learning process, the teacher needs to move among the groups and ask strategic questions to keep groups focused and moving ahead without prompting with too much information. The teacher also needs to be skillful in encouraging reluctant students to become part of the group effort. All members are encouraged to ask questions of each other. Often, one student is able to answer a question that another has, thus eliminating the need to ask the teacher. When no one in the group is able to answer a particular question, a group designee confers with the teacher and brings the answer back to the group.

The teacher has to maintain individual accountability while working in groups. To do so, a commonly used method is to assign a number (one–four) to each member. Once the problem has been solved, the teacher calls on student number *three* in each group to explain their solution. On another problem, the teacher may call on student number *one*. This reporting method means that each member of the group must be prepared to explain the group's solution.

Lastly, the teacher needs to make decisions about student accountability. Will students work on tests collaboratively or individually? Will a grade for the group, individual grades, or a combination of both be assigned?

We urge the users of this book to explore the various grouping strategies suggested for each lesson. Keep in mind the various ways of grouping students and ways of promoting full involvement of all participants in the group. We believe working collaboratively will promote greater understanding of the mathematical concepts presented in this book and enhance students' self-confidence in solving problems. These strategies will also promote the incorporation of the very vital set of *Common Core Standards for Mathematical Practice* (SMPs).

About the Authors

DAN DOLAN   Dan Dolan retired in January 2004 after 13 years as Associate Director and then Director of the *Project to Improve Mastery in Mathematics and Science* (PIMMS), a professional development program for teachers of mathematics and science at Wesleyan University in Connecticut. He taught mathematics and science in grades 6-12 for 21 years in CA, ID, and MT before going to the Montana State office where he served as the State Mathematics and Computer Education Specialist for 10 years. He has taught mathematics teacher education courses, authored numerous articles, and co-authored three books. He has presented numerous workshops and conference sessions throughout the country. He served on the Mathematical Sciences Education Board (MSEB), the Board of Directors of the National Council of Teachers of Mathematics (NCTM) and was a member of the Grades 5-8 writing group for the NCTM *Curriculum and Evaluation Standards for School Mathematics.* He has served on various NCTM committees and editorial panels for special NCTM publications, and he chaired the editorial panel for the *Student Math Notes*.

MARI MURI   Mari Muri retired in June 2003 after 15 years as Mathematics Consultant from the Connecticut State Department of Education. She actively continues her work with students, teachers, and administrators as a Senior Consultant for PIMMS at Wesleyan University. Prior to her state position, she was the Mathematics Instructional Consultant for the Killingly, CT Public Schools. She has been involved in education for 31 years and has taught at the elementary and university teacher education levels. She served on the MSEB, was a member of the writing team for the NCTM *Assessment Standards,* served as president of the Association of State Supervisors of Mathematics, was a member of the National Assessment of Educational Progress (NAEP) Math Standing Committee, served on the National Advisory Committee for the Center to Study Mathematics Curriculum, and as a member of the editorial panel for NCTM's *Mathematics Teaching in the Middle School.* Currently she serves as Chair of NCTM's Mathematics Education Board of Trustees. She recently completed her service on the Board of Directors for the National Council of Supervisors of Mathematics (NCSM) and the Board of Directors of the US Math Recovery Council. She remains current with the national state of mathematics education and is conversant with the Common Core State Standards for Mathematics.

JIM WILLIAMSON   In August 2007, Jim Williamson retired as an Adjunct Assistant Research Professor of Mathematics at the University of Montana in Missoula. From 2002 until his retirement, he supervised the writing team for the *Middle Grades MATH Thematics Phase II* grant, an NSF-funded project to revise the *Middle Grades MATH Thematics* curriculum. From 1998-2002, Jim was employed half time in the Mathematics Department, where he taught courses for prospective teachers, and half time on the *Show-Me Project,* an NSF-funded project supporting the dissemination and implementation of standards-based middle grades mathematics curricula. From 1992 to 1997, Jim chaired the writing team for the *Six Through Eight Mathematics (STEM)* middle school mathematics curriculum project. Before joining the STEM Project, Jim served for six months as the Interim Mathematics and Computer Education Specialist in the Montana Office of Public Instruction, as a Visiting Instructor of Mathematics at Montana State University for one year, and as the Mathematics Specialist for the Billings (MT) public schools for four years. He has been involved in mathematics education for over 40 years and has taught mathematics at all levels from fourth grade through university. He received a Presidential Award for Excellence in Mathematics Teaching in 1984. He has authored several articles, co-authored eight books, and presented staff development workshops and conference sessions at meetings throughout the country. He served on the NCTM committee that developed tests and coaching materials for the MATHCOUNTS competitions and chaired the committee for the 1989 competition.

Chapter 1
Thinking Critically

"A teacher of mathematics has a great opportunity. If the teacher fills the allotted time with drilling students with routine operations, the teacher kills their interest, hampers their intellectual development, and misuses the opportunity. But if the teacher challenges the curiosity of the students by setting them problems proportionate to their knowledge, and helps them solve their problems with stimulating questions, the teacher may give them a taste for, and some means of independent thinking."

George Polya
How to Solve It, 1957

"Problem solving means engaging in a task for which the solution method is not known in advance. In order to find a solution, students must draw on their knowledge, and through this process, they will often develop new mathematical understandings. Solving problems is not only a goal of learning mathematics but also a major means of doing so. Students should have frequent opportunities to formulate, grapple with, and solve complex problems that require a significant amount of effort and should then be encouraged to reflect on their thinking."

—Principles and Standards for School Mathematics

"Being able to reason is essential to understanding mathematics. By developing ideas, exploring phenomena, justifying results, and using mathematical conjectures in all content areas and—with different expectations of sophistication—at all grade levels, students should see and expect that mathematics makes sense. Building on the considerable reasoning skills that children bring to school, teachers can help students learn what mathematical reasoning entails."

—Principles and Standards for School Mathematics

The activities in this chapter are designed to help you improve your problem-solving, reasoning, and communication skills. Good problem solvers need to know a variety of techniques for solving problems, so the primary focus of the activities is on developing and applying a variety of problem-solving strategies: *look for a pattern, make a table, use logical reasoning, make a model*, and *elimination*.

As you complete the activities, you will learn to make conjectures based on observations and data, to verify and generalize the conjectures, and to communicate your results to others. This is the essence of mathematical inquiry. The problem-solving strategies and processes are the tools you will use throughout this book to explore and develop mathematical concepts.

1

**Correlation of Chapter 1 Activities to the
Common Core Standards of Mathematical Practice**

Activity Number and Title		Standards of Mathematical Practice
1:	When You Don't Know What to Do	SMP 1, SMP 4, SMP 5, SMP 7, SMP 8
2:	What's the Pattern?	SMP 4, SMP 7, SMP 8
3:	Fascinating Fibonacci	SMP 1, SMP 6, SMP 7
4:	What's the Rule?	SMP 4, SMP 7, SMP 8
5:	Ten People in a Canoe	SMP 1, SMP 4, SMP 7
6:	Magic Number Tricks	SMP 1, SMP 2, SMP 8
7:	An Ancient Game	SMP 1, SMP 8
8:	What's the Number?	SMP 1, SMP 4, SMP 7
9:	Eliminate the Impossible	SMP 1, SMP 7, SMP 8

Activity 1: When You Don't Know What to Do

PURPOSE	Introduce a four-step approach to solving problems.
COMMON CORE SMP	SMP 1, SMP 4, SMP 5, SMP 7, SMP 8
MATERIALS	Pouch: Colored Squares Online: Half-centimeter Graph Paper
GROUPING	Work individually or in pairs.
GETTING STARTED	Problem solving has been described as "what you do when you don't know what to do." As you investigate the following problem, you will learn a four-step approach that may help you solve problems.

Eighteen squares can be arranged into two congruent staircases in which no squares overlap and each step up contains exactly one less square than the step below it. One way this can be done is shown at the left.

Eight squares cannot be arranged into two staircases in this way.

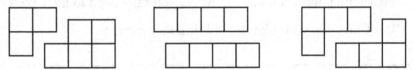

Non-congruent staircases   *Not staircases*   *Step up has two less squares than the one below*

What other numbers of squares can be arranged into two congruent staircases in this way?

UNDERSTAND THE PROBLEM

The first step in the four-step approach is to make sure you *understand the problem*. Among other things, this involves reading the problem carefully, sometimes several times, to be certain you understand what the question is. You must also identify the information needed to solve the problem and determine whether any of the information is missing.

1. Describe what is meant by a staircase in this problem.

2. What does it mean for shapes to be congruent?

3. a. Arrange 30 squares into two congruent staircases.

   b. How many different ways can you do it?

4. Restate the problem in your own words.

5. How did answering these questions help you understand the problem?

DEVISE A PLAN

Once you are sure you understand the problem, the next step is to *devise a plan* for solving it. This often involves choosing a problem-solving strategy or strategies to use in solving the problem.

1. Describe how the problem could be modeled using graph paper or squares.

2. What strategies other than *make a model* might be useful in solving the problem?

3. Describe how you would attempt to solve the problem.

CARRY OUT THE PLAN

Carrying out the plan involves using your chosen strategies to attempt to solve the problem. If a particular approach doesn't work, you may need to alter your plan.

1. Carry out your plan for solving the problem.

LOOK BACK

The final step is to *look back* over your solution not just to check your work, but to see what you have learned from solving the problem. Looking back involves checking that your answer is reasonable and that it answers the question that was asked. It also involves looking for other ways to solve the problem and looking for connections to other problems or mathematical ideas.

1. Does your solution include a way to test whether or not a given number of squares can be arranged into two congruent staircases? If not, try to find a way.

2. Try to verify your solution by solving the problem a different way.

3. a. How many different ways can 120 squares be arranged into two congruent staircases?

   b. How can you determine how many different ways a given number of squares can be arranged into congruent staircases?

4. a. How is the staircase problem related to finding the whole number factors of a number?

   b. How is it related to finding the odd and even factors?

Activity 2: What's the Pattern?

PURPOSE Identify and extend numerical and pictorial patterns and explore relationships between the terms of sequences and the term numbers.

COMMON CORE SMP SMP 4, SMP 7, SMP 8

GROUPING Work individually or in groups of 2 or 3.

GETTING STARTED Fill in the blanks with the numbers or pictures that complete the sequence, and briefly explain the rule you used. In some cases, an intermediate term or the last term is given so that you can check your work.

1.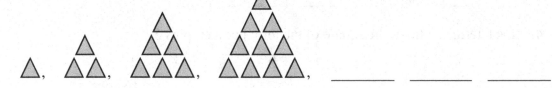

2.

3.

4.   3,   4,   3,   4,   5,   3,   4,   5, ___, ___, ___, ___, ___,   7

5.   2,   5,   8, 11, ___, ___, ___, ___, ___, 29

6. 53, 46, 39, 32, ___, ___, ___,   4, ___, ___

7.   1,   3,   6, 10, ___, ___, ___, ___, 55

8.   4,   7, 12, 19, ___, ___, ___, ___, ___, ___

9.   2,   4,   8, 16, ___, ___, ___, 256, ___, ___

10. 729, 243, 81, 27, ___, ___, ___, ___, ___, $\dfrac{1}{27}$

11.   3,   5,   8, 13, 21, ___, ___, ___, 144, ___, ___

12. Explain how the sequences in Exercises 1 and 5 and in Exercises 3 and 7 are related.

13. a. Explain how the sequences in Exercises 2 and 3 are related.

   b. How can the triangles in each term of the sequence in Exercise 3 be rearranged to illustrate this relationship?

Use the given rule to determine the first eight terms of each sequence.

Example: Each term is the term number times 4, plus 1.

First Term	Second Term	Third Term	Fourth Term
$1 \times 4 + 1$	$2 \times 4 + 1$	$3 \times 4 + 1$	$4 \times 4 + 1$
5	9	13	17

1. Each term is 5 times the term number, plus 2.

    1   2   3   4   5   6   7   8   9

    ____ 12 ____ ____ ____ ____ ____ ____ 47

2. Each term is 1 less than 3 times the term number.

    1   2   3   4   5   6   7   8   9

    ____ 5 ____ ____ ____ ____ ____ ____ 26

    Explain how this rule describes the shapes in the sequence in Exercise 1 on page 5.

3. Each term is the term number times –3, plus 47.

    1   2   3   4   5   6   7   8   9

    ____ ____ 38 ____ ____ ____ ____ ____ 20

4. Each term is 2 times the square of the term number, plus 5.

    1   2   3   4   5   6   7   8   9

    ____ ____ ____ ____ ____ ____ ____ ____ 167

5. Each term is the term number times the next term number.

    1   2   3   4   5   6   7   8   9

    ____ ____ ____ ____ ____ ____ ____ ____ 90

    Explain how this rule describes the shapes in the sequence in Exercise 2 on page 5.

6. Find the difference between successive terms in the sequences in Exercises 1–3. What do you notice? Does the same thing occur in Exercises 4 and 5? Explain.

7. If you did not know the rules for the sequences in Exercises 4 and 5, how could you find the next five terms of each sequence?

8. Find the missing terms in the following sequence

    2,   4,   8,   16,   30,   ____ ,   ____ ,   ____ ,   186,   ____

9. Compare the sequence in Exercise 8 to the sequence in Exercise 9 on page 5. Explain the difference between these two sequences.

Activity 3: Fascinating Fibonacci

PURPOSE Use a spreadsheet to explore patterns.

COMMON CORE SMP SMP 1, SMP 6, SMP 7

MATERIALS Other: A computer with spreadsheet software

GROUPING Work individually or with a partner.

GETTING STARTED Leonardo of Pisano (1170–1250), who is better known by the nickname Fibonacci, was one of the most talented mathematicians of the thirteenth century. His book *Liber abaci*, published in 1202, contained the fascinating, although somewhat unrealistic, problem about rabbit breeding paraphrased below.

Suppose a newborn pair of rabbits, a male and a female, is put in a field surrounded on all sides by a high wall. How many pairs will there be in one year if none of the rabbits die and, beginning at age 2 months, each pair produces another new pair every month?

1. To make sure you understand the rabbit problem, complete the following table.

Beginning of month	Number of newborn pairs	Number of 1-month-old pairs	Number of pairs 2 months old or older	Total number of pairs
1	1	0	0	1
2	0	1	0	1
3	1	0	1	2
4	1	1	1	3
5	2		2	
6				
7				

The original pair is now 1 month old.

The original pair is 2 months old and produces a new pair.

The original pair produces another new pair. The pair born in month 3 is now 1 month old.

The original pair and the pair born in month 3 are both 2 months old or older and produce new pairs.

2. a. From the second month on, how is the number of 1-month-old pairs related to the number of newborn pairs the preceding month?

   b. How can the number of 1-month-old pairs and the number of pairs 2 months old or older in any month be used to find the number of pairs 2 months old or older the next month?

   c. From the second month on, how is the number of newborn pairs each month related to the number of pairs 2 months old or older?

CREATING A SPREADSHEET

Spreadsheets were originally designed with business applications in mind, but they are also excellent problem-solving tools. Follow the steps below to use a spreadsheet to create a table like the one on the preceding page and to extend it to solve the rabbit problem.

1. a. Highlight cells A1 through E1 and select **Column Width** from the **Format** menu. Set the column width to 12 characters or 1.1 inches.

    b. Enter the column headings **Month, Newborn, 1 Month Old, 2 Months Old**, and **Total Pairs** in cells A1 through E1 respectively.

    c. Enter the numbers from the first row of the table in cells A2 through E2.

2. a. Since the number of each month is 1 more than the number of the preceding month, we can use a formula to generate the number of each month. Enter the formula "= A2 + 1" in cell A3. What happens?

    b. Highlight cells A3 through A14 and select **Fill Down** from the **Fill** menu. (If your spreadsheet does not have a **Fill Down** command, **Copy** cell A3 and **Paste** it into cells A4 through A14.) What happens?

    c. Why must the table go to 13 months?

3. The fact that from the second month on, the number of 1-month-old pairs is equal to the number of newborn pairs the preceding month can be used to complete column C.

    a. Select cell C3 and type "=". Click on cell B2 and press **Enter**. What happened?

    b. Highlight cells C3 through C14 and **Fill Down**.

4. For any month, the number of pairs 2 months old or older is the sum of the number of 1-month-old pairs and the number of pairs 2 months old or older the previous month. This relationship can be used to complete column D.

    a. Enter the formula "= C2 + D2" in cell D3. (Remember, instead of typing "C2" and "D2" you can simply click on the cells.)

    b. Highlight cells D3 through D14 and **Fill Down**.

5. The number of newborn pairs and the number of pairs 2 months old or older each month are equal. Use this fact to complete column B.

6. a. Enter the formula "= B3 + C3 + D3" in cell E3 to find the total number of pairs for month 2. (The formula can also be entered by highlighting cells B3 through E3 and clicking on the Σ (**Auto Sum**) in the tool bar.)

    b. Highlight cells E3 through E14 and **Fill Down** to complete Column E.

7. How many rabbit pairs will there be after one year?

8. The number pattern in column E is known as the *Fibonacci sequence*. The numbers in the sequence are called *Fibonacci numbers*. How are any two consecutive terms of the Fibonacci sequence used to find the next term in the sequence?

LOOKING FOR PATTERNS

The Fibonacci numbers have many interesting properties. Creating a spreadsheet can help you explore them.

1. a. Use the result from Exercise 8 on the preceding page to create a spreadsheet that contains the first 20 Fibonacci numbers in column A. Label the column **Fibonacci #s**.

    b. In column B, calculate the sums $1, 1 + 1, 1 + 1 + 2, 1 + 1 + 2 + 3, \ldots$ of the Fibonacci numbers. Label the column **Sums**.

    c. How are the sums in column B related to the Fibonacci numbers in column A?

    d. Check your conjecture in Part c by subtracting 1 from each Fibonacci number and entering the result in column C. Label the column **Fib. # – 1**.

2. a. In column D, calculate the square of each Fibonacci number. Label the column **Squares**.

    b. In column E, calculate the sums of the squares $1, 1 + 1, 1 + 1 + 4, 1 + 1 + 4 + 9, \ldots$ Label the column **Sum of Squares**.

    c. In column F, calculate the products $1 \times 1, 1 \times 2, 2 \times 3, 3 \times 5, 5 \times 8, \ldots$ of consecutive Fibonacci numbers. Label the column **Products**. How are columns E and F related?

    d. Make a conjecture about the sum of the squares of the first n Fibonacci numbers.

3. In column F, calculate the quotients $1 \div 1, 2 \div 1, 3 \div 2, 5 \div 3, 8 \div 5, \ldots$ of consecutive Fibonacci numbers. Label the column **Quotients**. What do you notice?

The number that the ratios of the consecutive Fibonacci numbers approach is the *Golden Ratio*. It arises in art, music, nature, and architecture. You will encounter the Fibonacci sequence and Fibonacci numbers again in later chapters.

Activity 4: What's the Rule?

PURPOSE	Develop a procedure for determining a rule that describes the general term of an arithmetic sequence.
COMMON CORE SMP	SMP 4, SMP 7, SMP 8
GROUPING	Work individually or in pairs.
GETTING STARTED	Fill in each blank to discover a method for determining the rule that generates an arithmetic sequence.

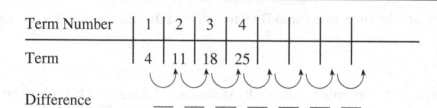

Term Number	1	2	3	4			
Term	4	11	18	25			

Difference ___ ___ ___ ___ ___ ___ ___

What is the constant difference? _____

Term Number		Constant Difference			What Was Done?		To Get	
1	×	7	→	7	_____	=	4	First Term
2	×	7	→	____	_____	=	11	Second Term
3	×	____	→	____	_____	=	____	Third Term
10	×	____	→	____	_____	=	____	Tenth Term
50	×	____	→	____	_____	=	____	Fiftieth Term

Write a sentence, like those in Activity 2, that states a rule for generating the terms in the sequence.

Use variables to write the rule as an equation.

For each of the following sequences:
• Fill in the missing numbers.
• Find a rule that generates the terms in the sequence.
• Determine the 25th and 100th terms of the sequence.

											Rule	25th Term	100th Term
1.	9,	13,	17,	21,	____,	____,	____,	____,	...		_____	____	____
2.	2,	9,	16,	23,	____,	____,	____,	____,	...		_____	____	____
3.	−3,	−1,	1,	3,	____,	____,	____,	____,	...		_____	____	____
4.	98,	96,	94,	92,	____,	____,	____,	____,	...		_____	____	____
5.	77,	74,	71,	68,	____,	____,	____,	____,	...		_____	____	____

Activity 5: Ten People in a Canoe

PURPOSE	Introduce the simplify problem-solving strategy and apply the make a table, make a model, and patterns strategies.
COMMON CORE SMP	SMP 1, SMP 4, SMP 7
MATERIALS	Pouch: Five each of two different-colored squares
GROUPING	Work individually or in groups of 2 or 3.

Ten people are fishing from a canoe. The seats in the canoe are just wide enough for one person to sit on, and the center seat is empty. The five people in the front of the canoe want to change seats and fish from the back of the canoe, and the five people in the back of the canoe want to fish from the front. Because the canoe is so narrow, only one person may move at a time. A person changing seats may move to the next empty seat, or step over one other person to reach an empty seat. Any other move will capsize the canoe.

What is the minimum number of moves needed to exchange the five people in the front with the five in the back?

HINT: Sometimes, the best approach to solving a problem is to simplify it by considering easier cases of the same problem. Use squares of two different colors to represent the people in the canoe and a model like the one below to represent the seats.

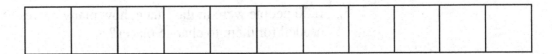

Simplify the problem by solving easier cases. The solution to the problem for two people in the canoe is shown below.

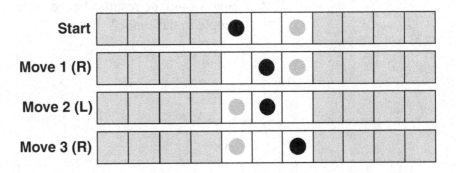

1. Solve the problem for four people. Record the results in the table below.

Number of People	2	4	6	8	10
Number of Pairs	1	2	3	4	5
Minimum Number of Moves	3				
Sequence of Moves	RLR				

2. Complete the table.

3. Look for two patterns, one for how the moves should be made and one for the minimum number of moves. Describe the patterns you found.

4. How many people in the canoe would produce the following sequence of moves?

   R LL RRR LLLL RRRRR LLLLLL RRRRRR LLLLLL RRRRR LLLL RRR LL R

5. If 30 people were in the canoe, how many moves would be needed for them to change places?

6. How would the results change if there was an odd number of people in the canoe?

EXTENSION   **The Legend of the Tower of Brahma**

It is said that in a temple at Benares, India, the priests work continuously moving golden disks from one diamond needle to another. It seems that when the world was created, the priests of Benares were given three diamond needles and 64 golden disks. The priests were told that they were to place the disks on one of the needles in decreasing order of size and then move the whole pile to one of the other two needles, moving only one disk at a time and never placing a larger disk on top of a smaller one. According to the legend, God told the priests, "When you finish moving the pile, the world will end."

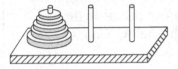

We can simulate the priests' problem by using coins, Cuisenaire rods, or different-sized squares cut from paper to represent the disks. Each peg can be represented by a square in a model like the one below.

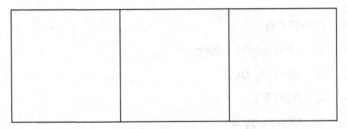

Stack the objects in one of the squares in decreasing order of size. The goal is to move the stack of objects from one square to another in the fewest possible moves. There are two rules: (a) only one object may be moved at a time, and (b) a larger object may never be placed on top of a smaller one.

1. What is the minimum number of moves required to move five objects from one square to another? **HINT:** Look for two patterns, as in the previous problem.

2. Suppose the priests move one disk every second without stopping. How long will it take them to move:

   a. 10 disks?   b. 30 disks?

   c. 50 disks?   d. all 64 disks?

Activity 6: Magic Number Tricks

PURPOSE	Translate verbal phrases into algebraic expressions and reinforce the work backward problem-solving strategy.
COMMON CORE SMP	SMP 1, SMP 2, SMP 8
MATERIALS	Other: Calculator
GROUPING	Work individually.

1. Dr. Wonderful, the Mathematical Magician, astounds crowds with his amazing ability to read people's minds. Here are the directions that he gives to five people in the crowd. Follow Dr. Wonderful's directions and complete the table below. Choose a different number for each person.

PERSON	1	2	3	4	5
1. Pick any number.	_____	_____	_____	_____	_____
2. Multiply by 3.	_____	_____	_____	_____	_____
3. Add 30.	_____	_____	_____	_____	_____
4. Divide by 3.	_____	_____	_____	_____	_____
5. Subtract your original number.	_____	_____	_____	_____	_____

2. Dr. Wonderful tells the people to write their answers on a sheet of paper but not to reveal them to anyone else. He closes his eyes, concentrates deeply, and then claims that he knows each person's answer. Suppose you picked 239. What do you think Dr. Wonderful would say your answer is? Why?

3. To learn why the trick works, let n be the number. Write an algebraic expression for each step and record it below.

   Step 1:

   Step 2:

   Step 3:

   Step 4:

   Step 5:

   The result is _____.

People in the crowd plead with Dr. Wonderful to teach them a magic trick that they can use with their friends. He agrees and explains the following.

	Example	**Your Number**
1. "Pick a number." (Begin with numbers less than 15.)	4	_____
2. "Multiply by three."	12	_____
3. "Add seven to the product."	19	_____

Dr. Wonderful says, "If you tell me your final answer, I will tell you your original number." One person says, "19."

Dr. Wonderful closes his eyes and thinks deeply.

1. He subtracts 7.	$19 - 7 = 12$	_____
2. He divides by 3 and announces the answer.	$12 \div 3 = 4$	_____

"Try again with a harder one," calls the crowd.

1. "Pick a number," he says	7	_____
2. "Add five."	12	_____
3. "Multiply the sum by four."	48	_____
4. "Subtract seven."	41	_____

Dr. Wonderful's steps can be described by the algebraic equation $4(x + 5) - 7 = $ Answer.

In order to determine the starting number, Dr. Wonderful *works backward* to undo the equation. For each step, he does the operation that is opposite to the one he stated.

1. Begin with 41.	41	_____
2. Add seven.	48	_____
3. Divide by four.	12	_____
4. Subtract five.	7	_____

Work in groups. Have each person in the group play the part of Dr. Wonderful who will think of a set of directions for a magic trick. Dr. Wonderful will tell the members of the group to think of a number and then read the steps to be followed.

When all have determined their answers, Dr. Wonderful will "read their minds," announcing the starting number for each person in the group.

1. One of Dr. Wonderful's other mind-reading tricks involves birthdays. Use a calculator and follow along with the crowd as he gives the directions. Press ⟨=⟩ on your calculator after each step.

    1. Enter the month of your birthday.

    2. Multiply by 5.

    3. Add 20.

    4. Multiply by 4.

    5. Subtract 7.

    6. Multiply by 5.

    7. Add the day of your birthday.

    8. Subtract the number of days in a non-leap year.

2. "Oh," "Ah," and "Look at that" can be heard throughout the crowd. What do people see on the display of their calculator?

3. To the right of each step in Exercise 1, write an algebraic expression that correctly describes Dr. Wonderful's direction. Study the sequence of expressions and then explain how place value helps to explain how this "magic number trick" works.

EXTENSION   Write some magic tricks of your own. Write the algebraic expression for each step so that you can justify your final result. Try them out on your classmates.

Activity 7: An Ancient Game

PURPOSE	Introduce the work backward problem-solving strategy.
COMMON CORE SMP	SMP 1, SMP 8
MATERIALS	Other: One calculator
GROUPING	Work in pairs.
GETTING STARTED	This is a version of a game called NIM. It is a game for two players. Beginning with 17, the players alternate turns subtracting 1, 2, or 3 from the number on the calculator display. The player who makes the display read 0 is the winner.

SAMPLE GAME

Player	Keys Pressed	Display
A	17 − 3 =	14
B	− 2 =	12
A	− 3 =	9
B	− 1 =	8
A	− 3 =	5
B	− 3 =	2
A	− 2 =	0

Player A wins!

1. Play the game several times.

2. a. In the Sample Game, what number could Player B have subtracted on his/her last turn and been sure to win the game? Explain.

   b. Find a strategy for winning the game.

3. There are many variations of this game.

   a. Try starting with 25 and subtracting 1, 2, 3, or 4. How does this change the strategy for winning the game?

   b. What is the strategy for winning the following version of NIM? Start with 47 and alternate turns subtracting 3, 5, or 7 from the number on the display. The first player to get a number less than or equal to 0 on the display is the winner.

Activity 8: What's the Number?

PURPOSE	Apply the elimination problem-solving strategy.
COMMON CORE SMP	SMP 1, SMP 4, SMP 7
GROUPING	Work individually or in groups of 3 or 4.
GETTING STARTED	Use the process of elimination to solve the following number puzzles.

1. Circle the number below that is described by the following clues. Keep a record of the order in which you use the clues.

    a. The sum of the digits is 14.

    b. The number is a multiple of 5.

    c. The number is in the thousands.

    d. The number is not odd.

    e. The number is less than 2411.

1580	2660	2570	1922	905
2290	1355	1058	1455	770
	1832	2435	1770	1680
		860		

2. What clue or combination of clues did you use first? Why?

---

1. Solve the following number riddle.
    - I am a positive integer.
    - All my digits are odd.
    - I am equal to the sum of the cubes of my digits.
    - I am less than 300.

    Who am I? _____

2. In what order did you use the clues? Why?

Rebecca has a collection of basketball cards. When she puts them in piles of two, she has one card left over. When she puts them in piles of three or four, there is also one card left over, but when she puts them in piles of five there are no cards left over.

If Rebecca has fewer than 100 basketball cards, what are the possible numbers of cards she could have?

Activity 9: Eliminate the Impossible

PURPOSE Introduce the method of indirect reasoning.

COMMON CORE SMP SMP 1, SMP 7, SMP 8

GROUPING Work individually or in groups of 2 or 3.

Andrea was visiting her Uncle Ralph, who has a large gumball collection. When she asked if she could have some, he said yes, but only if she could solve a problem for him. He told her that he has three jars, each covered so that no one can see the color of the gumballs. One jar is labeled red, the second green, and the third red-green. However, he said, no jar has the correct label on it. She could reach into one jar and take one gumball. Then she had to tell him the correct color of the gumballs in each jar. She reached into the jar labeled red-green and pulled out a red gumball.

1. Are there any green gumballs in that jar? Why?

2. What is the correct label for the jar labeled red-green? Explain your answer.

3. Can the jar labeled red contain red and green gumballs? Why?

4. What are the correct labels for each of the jars?

Jorge claims that he has a certain combination of U.S. coins but he cannot make change for a dollar, half dollar, quarter, dime, or nickel. Is this possible? If so, what is the greatest amount of money Jorge could have, and what coins would they be? He does not have any dollar coins.

Greatest amount: _____

Coins: _____

Students in the fifth grade were playing a trivia game involving states, state birds, and state flowers. They knew that in Alaska, Alabama, Oklahoma, and Minnesota, the state flowers are the camellia, forget-me-not, pink-and-white lady's slipper, and mistletoe. The state birds are the common loon, yellowhammer, willow ptarmigan, and scissor-tailed flycatcher. No one knew which bird or flower matched which state. They called the library and received the following clues. Use the clues to complete the table below.

a. The flycatcher loves to nest in the mistletoe.

b. The forget-me-not is from the northernmost state.

c. Loons and lady's slippers go together, but Minnesota and mistletoe do not.

d. The yellowhammer is from a southeastern state.

e. The willow ptarmigan is not from the camellia state.

State				
Flower				
Bird				

Which of the clues were the key(s) to solving the puzzle? Explain your reasoning.

Each year, the Calaveras County Frog Jumping Contest is held at Angel's Camp, California. In last year's contest, four large bullfrogs—Flying Freddie, Sailing Susie, Jumping Joe, and Leaping Liz—captured the first four places. Each frog was decorated with a brightly colored bow before the competition began. From the following clues, determine which frog won each place and the color of its bow.

a. Joe placed next to the frog with the purple bow.

b. The frog with the yellow bow won, and the frog with the purple bow was second.

c. The colors on Freddie's bow and Susie's bow mix to form orange.

d. The color of the remaining bow was green.

Construct a table similar to the one above to help organize your work.

Chapter Summary

In this chapter, you studied some of the tools and processes used to explore and develop mathematical concepts. The goal was to help you develop your own problem-solving ability.

In Activity 1, you explored a four-step approach to problem solving. One advantage of this approach is that it gives you a way to get started on solving a problem. Many of the remaining activities developed problem-solving strategies, such as *look for a pattern*, *simplify*, and *work backward*, which are often used in conjunction with the four-step approach.

The study of patterns and functions is a central theme in mathematics. In this chapter, you learned various ways to analyze patterns. In Activity 2, you learned to recognize different types of patterns.

a. Patterns like the ones in Exercise 4 that have a growing core:

   34 345 3456 34567 . . .

b. Patterns, such as those in Exercise 3 and Exercise 9, which grow in a predictable way:

   •
   •, • •, • • •,   . . .

   2, 4, 8, 16, . . .

c. Patterns, such as those in Exercises 5 and 10, that can be extended by adding or subtracting a constant or by multiplying or dividing by a constant:

   2, 5, 8, 11, . . .   and   729, 243, 81, 27, . . .

d. Patterns like the one in Exercise 8 where there is a pattern in the differences between successive terms:

   4, 7, 12, 19, . . .

While studying patterns, you learned some new terminology: *term*, *term number*, and *sequence*. You also learned that for many sequences, there is a rule that relates any term of the sequence to its term number.

One method you learned for analyzing and extending numeric sequences was to examine the differences between the successive terms of the sequence. In Activity 4, you found that in the cases where the differences were constant, you could use the difference to generate a rule for the general term of the sequence.

The methods you used to analyze and extend patterns based on your observations are examples of *inductive reasoning*. You discovered the limitations of the inductive reasoning process in Activity 2. No matter how many initial terms of a sequence you may know, there is generally more than one way to extend it.

Activity 3 explored the use of technology to communicate mathematically. In it, you investigated a classic problem posed by Leonardo of Pisano and discovered some fascinating properties of the Fibonacci numbers.

Activity 5 integrated many problem-solving techniques. You learned to simulate problems that could not be experienced firsthand, to apply the patterns strategy, and to use tables as an organizer. You also learned a new problem-solving strategy, *simplify the problem*—begin with a simple case of the problem and work through successively more complex cases until a general method of solution is discovered. This technique will be used extensively for investigating new mathematical concepts.

In Activity 6, you investigated some classic number tricks. As you analyzed the tricks, you discovered how using variables to translate the instructions into algebraic expressions could help explain the "magic."

The *work backward* problem-solving strategy was introduced in Activity 7 and the *logical reasoning* strategy was developed in Activities 8 and 9. Most of the activities began with a set of clues, or premises, that were accepted as true. By reasoning logically from these premises, you were able to conclude something about a number or situation. This process of deriving a conclusion by reasoning logically from a set of known premises is called *deductive reasoning*.

Usually, you were able to reason *directly* from the premises to the conclusion. However, in Activity 9, you had to test possible solutions by assuming they were true. If the assumption led to a contradiction of a known fact, then you knew that the proposed solution was not correct. This method, which was introduced through elimination, is known as *indirect reasoning*—it is used extensively in mathematics.

Logical reasoning is the cornerstone upon which mathematics is built. New mathematics is often discovered via inductive reasoning. But before a conjecture arrived at inductively is accepted as a fact, it must first be verified using deductive reasoning. It is this standard of proof that distinguishes mathematics from the other sciences.

The activities in this chapter were intended to provide only an informal introduction to inductive and deductive reasoning. You will learn more about these techniques later in this book and use them throughout it.

Chapter 2
Sets and Whole Numbers

"… understanding number and operations, developing number sense, and gaining fluency in arithmetic computation form the core of mathematics education for elementary grades. … As students gain understanding of numbers and how to represent them, they have a foundation for understanding relationships among numbers."
— *Principles and Standards for School Mathematics*

"Students understand the meanings of multiplication and division of whole numbers through the use of representations (e.g., equal-sized groups, arrays, area models, and equal 'jumps' on number lines for multiplication, and successive subtraction, partitioning, and sharing for division)."
— *Curriculum Focal Points for Prekindergarten through Grade 8 Mathematics*

The activities in this chapter make connections among many of the NCTM Standards. Using the concepts associated with sets (well-defined groups of objects), identifying the properties of the elements of a set, and sorting the elements into subsets according to specific properties are all fundamental processes of mathematics.

Some of the activities will engage you in sorting and classifying objects, describing their properties (attributes), and describing their similarities and differences. These activities promote reasoning and communication in mathematics. Explaining your reasoning, defending your conjectures, and evaluating input from others all promote greater understanding of mathematical concepts.

When engaged in problem-solving situations that involve sorting, classifying, and discriminating, reflect on the connections between these processes in science and mathematics. You will find that the direct and indirect reasoning skills you develop in this chapter will be an important asset in other problem settings throughout the book.

The number activities in this chapter are designed to help you develop a strong sense of number and an understanding of the basic operations of multiplication and division.

23

Correlation of Chapter 2 Activities to the
Common Core Standards of Mathematical Practice

Activity Number and Title		Standards of Mathematical Practice
1:	What's in the Loop?	SMP 1, SMP 2, SMP 3, SMP 4, SMP 6, SMP 7
2:	Loop de Loops	SMP 1, SMP 2, SMP 3, SMP 5, SMP 6, SMP 7
3:	Odd and Even Patterns	SMP 4, SMP 5, SMP 7, SMP 8
4:	Multiplication Arrays	SMP 6, SMP 7, SMP 8
5:	How Many Cookies?	SMP 6, SMP 7, SMP 8
6:	Find the Missing Factor	SMP 4, SMP 6, SMP 7
7:	Paper Powers	SMP 4, SMP 6, SMP 7
8:	The King's Problem	SMP 1, SMP 4, SMP 5, SMP 6, SMP 7

Activity 1: What's in the Loop?

PURPOSE   Use the elimination and logical reasoning problem-solving strategies to develop the concept of a set, to determine the attribute defining a set, and to explore the concepts of complement, equivalent sets, equal sets, and cardinality.

COMMON CORE SMP   SMP 1, SMP 2, SMP 3, SMP 4, SMP 6, SMP 7

MATERIALS   Pouch: Attribute Pieces
Online: Label Cards for Attribute Pieces
Other: Large loop of string

GROUPING   Work in pairs or in teams of two students each.

1. Place the loop between the players. Put the **RED** label card face up on the loop. Take turns placing pieces in the appropriate region, either inside the loop or outside it.

2. The pieces that are in the loop are the set of red attribute pieces. A *set* is any collection of objects. Each attribute piece that is in the loop is an *element* of the set of red attribute pieces. The *universal set* is the set that contains all the elements being considered in a given situation, in this case, the set of all attribute pieces.

    a. Use abbreviations to list the elements in the set of red attribute pieces. For example, use SRT for the small red triangle.

    b. The number of elements in a set is the *cardinality* of the set. What is the cardinality of the set of red attribute pieces?

    c. What is the cardinality of the universal set?

3. a. The *complement* of a set *A* is the set of all elements of the universal set that are not in set *A*. List the elements in the complement of the set of red attribute pieces.

    b. Where are these pieces in relation to the loop?

4. a. Two sets that contain the same number of elements are *equivalent*. Name a set of attribute pieces that is equivalent to the set of red pieces.

    b. Without actually counting the number of pieces in the two sets, how could you show that the sets are equivalent?

    c. Two sets are *equal* if and only if they contain exactly the same elements. Name a set of attribute pieces that is equal to the set of small attribute pieces.

5. Place the loop between the players. Shuffle the label cards and place them face down on the table. Player A picks a card from the pile, looks at it, and places it face down next to the loop without showing it to the other player. Player B chooses an attribute piece and places it in the loop. Player A then tells whether the placement is correct or not based on the label card. Play continues until Player B can correctly name the set by identifying the exact attribute on the label card. Players then switch roles.

Activity 2: Loop de Loops

PURPOSE   Use indirect reasoning to determine the attribute defining a set, and to explore the concepts of union, intersection, empty set, and subset.

COMMON CORE SMP   SMP 1, SMP 2, SMP 3, SMP 5, SMP 6, SMP 7

MATERIALS   Pouch: Attribute Pieces
Online: Label Cards for Attribute Pieces
Other: Two large loops of string

GROUPING   Work in pairs or in teams of two students each.

1. a. Place the loops between the players as shown in the diagram below. Put the LARGE label card face up on one loop and the RED label card face up on the other. Take turns placing pieces in the appropriate loop or outside the loops.

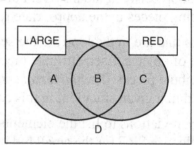

   b. Why must the loops be overlapped in order to place the attribute pieces correctly?

2. a. Complete the following sentence describing the attribute pieces that are in the shaded region.

   The shaded region is the set of all attribute pieces that are

   _____ .

   b. The set in Part a is the *union* of the set of red pieces and the set of large pieces. How many pieces are in the union of the two sets, that is, how many pieces are either RED or LARGE?

3. a. Complete the following sentence describing the attribute pieces that are in region B.

   Region B is the set of all attribute pieces that are

   _____ .

   b. The set in Part a is the *intersection* of the set of red pieces and the set of large pieces. How many pieces are in the intersection of the two sets, that is, how many pieces are both RED and LARGE?

4. The *empty set* is a set that contains no elements. What labels could be used for the loops so that the intersection of the two loops is empty?

5. Set *A* is a subset of set *B* if and only if every element of *A* is also an element of *B*. What labels could be used for the two loops so that one loop would be a subset of the other?

6. a. How many pieces are LARGE but not RED?

   b. What region contains the pieces that are LARGE but not RED?

7. A few of the attribute pieces have been placed in the loops in the diagram below. What are the labels for the loops? How do you know?

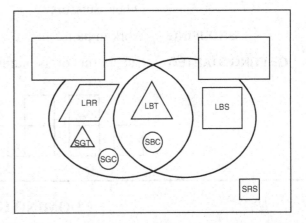

8. Place the loops between the players. Overlap them as shown below. Shuffle the label cards and place them face down on the table. Player A picks two label cards and, without showing them to the other player, places them face down, one on each loop as shown.

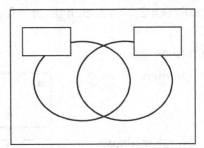

Player B chooses an attribute piece and places it in one of the four regions. Player A then indicates whether the placement is correct or not according to the labels on the cards that have been placed on the loops. If the placement is incorrect, the first student may try another region, or put the piece back in the pile and try another attribute piece. Play continues until Player B can correctly identify both label cards. Players then switch roles.

9. Repeat Exercise 8 using three loops and three label cards.

Activity 3: Odd and Even Patterns

PURPOSE	Explore the sums of odd and even numbers and discover the relationship between the two addends and the sum.
COMMON CORE SMP	SMP 4, SMP 5, SMP 7, SMP 8
MATERIALS	Pouch: One set of Double-Six Dominoes Online: Centimeter Graph Paper Other: Scissors
GROUPING	Work in pairs.
GETTING STARTED	Cut six sets of the number shapes shown below from graph paper.

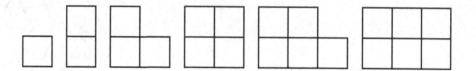

DOMINO GAME

- Remove the dominoes that have a blank. Place the remaining dominoes face down and mix well.
- Place the number shapes between the players.
- Player A selects a domino and the number shapes that match and then puts them together so that the result is a 2 × *n* rectangle or a figure that is as close to a 2 × *n* rectangle as possible.
- Record each turn as shown.

   Example: Domino

Number Shapes				
Equation	2	+	5	=	7
Odd/Even	**Even +**	**Odd**	**=**	**Odd**

- Players alternate turns. A player loses a turn when a pair of shapes cannot be found to match the domino.
- The game ends when neither player can make a match. The winner is the player with the greatest number of EVEN sums.

1. When you play the game, are you more likely to form an odd sum or even sum? Explain why.

2. Any even number can be written as $2n$, where n is a whole number. Suppose a set of dominoes had up to 100 dots on one side. If you cut out shapes for the even numbers, describe the even number shapes and how each shape is related to this algebraic expression.

3. If any even number can be written as $2n$, then any odd number can be written as _____ or _____, where n is a whole number.

4. If you cut out shapes for the odd numbers, describe the odd number shapes and how they relate to the algebraic expressions in Exercise 3.

5. Adding two even numbers can be represented as $2n + 2m = 2(n + m)$, where m and n are both whole numbers. Describe the relationship between the shapes of the two addends and the shape of the sum in this equation.

6. Repeat Exercise 5 for the addition of two odd numbers and for the addition of an odd and an even number.

Activity 4: Multiplication Arrays

PURPOSE	Develop the concept of multiplication as repeated addition and the commutative property for multiplication.
COMMON CORE SMP	SMP 6, SMP 7, SMP 8
MATERIALS	Pouch: Colored Squares
	Online: Centimeter Graph Paper and a Multiplication and Division Frame
	Other: Paper for recording
GROUPING	Work individually.
GETTING STARTED	In a multiplication problem, two factors determine a product. Every multiplication fact can be illustrated as a rectangular array on a multiplication and division frame. The factor to the left of the frame determines the number of groups (rows). The number above the frame determines the number in each group (columns).

Use the multiplication and division frame to construct each of the first 10 multiples of 4. Record your work on paper as shown below.

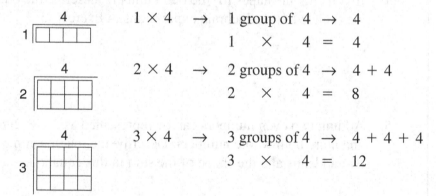

1. What happens to the rectangular arrays and the products in the Example when the same factors are used but their order is reversed, that is, 4 × 1, 4 × 2, 4 × 3?

2. On square grid paper, construct a 3 × 4 rectangle and a 4 × 3 rectangle. Cut out the rectangles and place one on top of the other. What must be done to align them so that they match each other? What can you conclude from this?

Activity 5: How Many Cookies?

PURPOSE	Develop the concept of division as repeated subtraction.
COMMON CORE SMP	SMP 6, SMP 7, SMP 8
MATERIALS	Pouch: Colored Squares Other: Paper for recording
GROUPING	Work individually or in pairs.
GETTING STARTED	The following is an example of the repeated subtraction model of division: "Tyler has 24 candies in a jar and wishes to decorate each cookie with six candies. How many cookies will he be able to decorate?" Solving $24 \div 6$ or $6\overline{)24}$, is illustrated below.

Example:

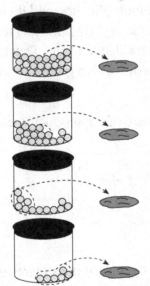

$$6\overline{)24}$$
$$\underline{-\ 6}$$
$$18$$
Remove 6 one time.

$$\underline{-\ 6}$$
$$12$$
Remove 6 a second time.

$$\underline{-\ 6}$$
$$6$$
Remove 6 a third time.

$$\underline{-\ 6}$$
$$0$$
Remove 6 a fourth time.

$$24 \div 6 = 4$$

---

1. Write five additional word problems that illustrate the repeated subtraction model for division and solve them as shown above.

2. Solve the following problems using the *repeated subtraction* feature on a four-function calculator.

   Example: $85 \div 17$

   $\boxed{8}\,\boxed{5}\,\boxed{-}\,\boxed{1}\,\boxed{7}\,\boxed{=}\,\boxed{=}\,\boxed{=}\,\boxed{=}\,\boxed{=}$

   $85 \div 17 = 5$

a. $96 \div 12$	b. $318 \div 37$	c. $107 \div 24$	d. $803 \div 73$
e. $94 \div 7$	f. $152 \div 19$	g. $297 \div 14$	h. $429 \div 33$

3. Explain how you determined the quotient and the remainder in each problem using the *repeated subtraction* model.

Activity 6: Find the Missing Factor

PURPOSE	Develop the concept of division as equal sharing (*partitioning*).
COMMON CORE SMP	SMP 4, SMP 6, SMP 7
MATERIALS	Pouch: Colored Squares Online: Multiplication and Division Frame Other: One die
GROUPING	Work individually or in pairs.
GETTING STARTED	The problem "Tasha has 15 marbles. She wishes to give the same number of marbles to each of five friends. How many marbles does each person receive?" is an example of the *partitioning* model for division. Use the multiplication and division frame to solve $15 \div 5$ or $5\overline{)15}$. Record the factor 5 (the divisor) to the left of the frame and place 15 squares inside the frame. Determine how many squares must be placed in each of the five groups to form a rectangular array.

Example: a.   Place one square of the product in each group.

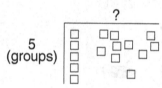

b.   Now arrange the rest of the squares into groups to construct a rectangular array.

?

5
(groups)

c.   How many squares are there in each group (row)? _____

?

5
(groups) $15 \div 5 = 3$

Student A rolls a die to determine a factor. Write the number to the left of the frame. Student B picks a handful of squares (without counting) to determine the product (dividend), and then uses the factor to construct a rectangular array as in the previous example.

Example: The number of squares in the handful equals 15.

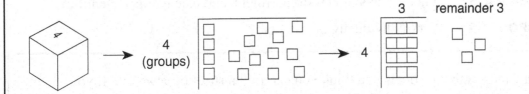

1. a. How many squares are in each of the 4 groups in the example? _____

    b. How many squares are remaining? _____

    c. Record your answer. 15 ÷ 4 = _____ R _____

2. Repeat this activity 10 times, alternating turns.

3. Describe the relationship between the *partitioning* model for division and multiplication.

EXTENSIONS   1. If you use the ⊕ key, 15 ÷ 4 = 3.75. How would you use the calculator to determine the *whole number* remainder? Show the sequence of keys to be pressed.

2. Use the ⊕ key on your calculator to find the quotient and whole number remainder in the following problems:

a. 64 ÷ 3	b. 42 ÷ 2	c. 116 ÷ 6
d. 57 ÷ 4	e. 94 ÷ 6	f. 43 ÷ 8
g. 81 ÷ 9	h. 79 ÷ 5	i. 157 ÷ 7

Activity 7: Paper Powers

PURPOSE	Develop an understanding of exponents and exponential change.
COMMON CORE SMP	SMP 4, SMP 6, SMP 7
MATERIALS	Other: Sheets of newsprint, rulers, and calculators (scientific)
GROUPING	Work individually or in pairs

1. Estimate the number of times you think you can fold a sheet of newsprint if you continue to fold the result in half each time. _____

2. Now, fold the sheet in half as many times as you can. After each fold, count the number of layers of paper and record the result for the **Number of Layers** in **Standard Form** in the table. What happens to the number of layers after each fold?

3. Starting with two folds, record the **Number of Layers** in **Factored Form** and in **Exponential Form** in the table.

Folds	0	1	2	3	4	5	6	7	8
No. Layers (Std. Form)	1	2							
No. Layers (Fact. Form)			2×2						
No. Layers (Exp. Form)				2^3					
Approx. Height (cm)								1.0	

4. How does your estimate for the number of folds in Exercise 1 compare to the actual number of folds you were able to make?

5. Examine the pattern of entries as you go from *right to left* in the Exp. Form row. If the pattern continues, what will be the correct entries for 1 fold? _____ 0 folds? _____

6. If a large sheet of newsprint is folded seven times as described above, the thickness is approximately 1.0 cm. Use this number to determine other entries for **Height of the Stack** in the table.

7. Use your calculator to extend the table and determine the number of layers needed to approximate your height.

   Your height (cm) _____ No. of layers _____ Height (from table) _____

8. a. If a sheet could be folded 30 times, would the stack reach the top of the Willis (formerly Sears) Tower, _____ an orbiting satellite, _____ the moon? _____

      Y/N   Y/N   Y/N

   b. Use the pattern in the table to determine the height after 30 folds. Describe how your answer compares to the height of the Willis Tower, the distance to an orbiting satellite, and the distance from Earth to the moon.

Activity 8: The King's Problem

PURPOSE   Apply problem-solving strategies to develop an understanding of exponential growth.

COMMON CORE SMP   SMP 1, SMP 4, SMP 5, SMP 6, SMP 7

MATERIALS   Other: Rice, measuring tools, and calculators (scientific)

GROUPING   Work individually or in groups of 2 or 3.

GETTING STARTED   Legend has it that when the inventor of the game of chess explained the game to his king, the king was so delighted he asked the man what gift he would like as a reward.

"My wants are simple," the man replied. "If you but give me one grain of rice for the first square on the playing board, two for the second, four for the third, and so on for all sixty-four squares, doubling the number of grains each time, I will be satisfied."

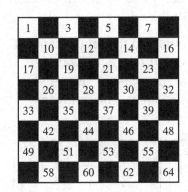

1. Suppose the king agreed to the request.

    a. How many grains of rice would the inventor receive?
Hint: How would the number of grains of rice on the seventh square compare to the total number of grains on the first six squares?

    b. How would the total number of grains of rice on the black squares compare to the total number of grains on the white squares?

2. a. How much would the number of grains of rice you found in Exercise 1(a) weigh?

    b. How many bushels of rice would this be? Explain how you got your answer.

    c. How large would a building need to be to hold the rice? (Make a sketch of the building and label its dimensions.)

    d. At today's prices, what would the retail value of the rice be?

3. Consult an almanac or the Internet to answer the following.

    a. Does the United States produce enough rice in one year to satisfy the inventor's request? Explain.

    b. Is enough rice produced in the world in one year to satisfy the inventor's request? Explain.

    c. How long would it take to produce the needed rice?

Chapter Summary

Sets and the operations on them were introduced in this chapter. The fields of mathematics, such as arithmetic, geometry, and algebra, are characterized by the study of particular sets of objects. This study entails an analysis of the elements of the set, their attributes, and what makes the set uniquely different from other sets.

In the first two activities, you sorted the elements of a set, the *universal set*, according to certain attributes. In Activity 1, when you made a choice to place a particular attribute piece in the loop, you guessed (made an assumption) that a certain label belonged on that loop. Given a YES answer, you gained information about the correct label. Given a NO answer, you gained information about which label was NOT correct, thus eliminating some labels, and reducing the number of pieces to try in order to complete the problem. Activity 2 extended this use of the elimination problem-solving strategy and the use of indirect reasoning to determine the proper labels for the sets.

As you completed the activities, you encountered the set concepts of *union, intersection, subset, complement, equal sets, equivalent sets,* and the *empty set*. The word OR described the *union* of sets, AND, the *intersection* of sets, and NOT, the *complement*. The following diagrams illustrate these concepts.

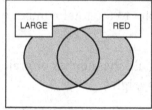   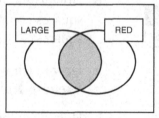   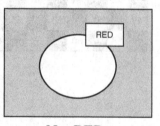

RED or LARGE	RED and LARGE	Not RED
Union	*Intersection*	*Complement*

Activity 3 introduced addition as the union of sets by combining (adding) *odd* and *even* numbers. After exploring the patterns generated by this addition, you found algebraic expressions to generalize the additions.

Activity 4 introduced multiplication of whole numbers as *repeated addition* and Activities 5 and 6 developed the meaning of division as *partitioning* and as *repeated subtraction*. These activities emphasized the inverse relationship between multiplication and division.

Activities 7 and 8 used the patterns problem-solving strategy to introduce exponents and exponential growth.

Chapter 3
Numeration and Computation

Give a man a fish, and he eats for a day. Teach him how to fish, and he eats for a lifetime.

" ... students should attain rich understanding of numbers—what they are; how they are represented with objects, numerals or on number lines; how they are related to one another; how numbers are embedded in systems that have structures and properties; and how to use numbers and operations to solve problems.... Representing numbers with various physical materials should be a major part of mathematics instruction in the elementary school grades."

—*Principles and Standards for School Mathematics*

The development of number sense, operation sense, place-value concepts, an understanding of the algorithms for the basic operations, and estimation strategies are among the most important tasks in teaching elementary mathematics. In Chapter 2, you began developing number sense ideas and refined your understanding of operations. This chapter continues the emphasis on these concepts and makes connections among concrete models, numbers, and algorithms for the basic operations. Through these models and representations, the elementary teacher promotes the development of number and operation sense.

This conceptual approach to teaching mathematics allows students to construct their own ideas of number and computation. The use of manipulatives provides a problem-solving environment for learning in which students are constantly discussing, conjecturing, visualizing, asking new questions as they review their work, and communicating with others about their findings.

The activities in this chapter provide interesting and challenging situations to increase your number sense and your understanding of place value and algorithms for operations. Each one provides an opportunity to explore mathematical concepts in a hands-on, problem-solving setting or in a game format.

Correlation of Chapter 3 Activities to the
Common Core Standards of Mathematical Practice

Activity Number and Title	Standards of Mathematical Practice
1: Regrouping Numbers	SMP 3, SMP 4, SMP 5, SMP 7
2: Find the Missing Numbers	SMP 1, SMP 2, SMP 4, SMP 5, SMP 6
3: It All Adds Up	SMP 4, SMP 5, SMP 6, SMP 7
4: What's the Difference?	SMP 4, SMP 5, SMP 6, SMP 7
5: Multi-digit Multiplication	SMP 4, SMP 5, SMP 6, SMP 7
6: Least and Greatest	SMP 1, SMP 3, SMP 6, SMP 7, SMP 8
7: Multi-digit Division	SMP 4, SMP 5, SMP 6, SMP 7
8: Target Number	SMP 1, SMP 4, SMP 5
9: A Visit to Fouria	SMP 1, SMP 2, SMP 4, SMP 5, SMP 6

Activity 1: Regrouping Numbers

PURPOSE   Develop number sense and an understanding of place-value and regrouping concepts by constructing models of numbers.

COMMON CORE SMP   SMP 3, SMP 4, SMP 5, SMP 7

MATERIALS   Pouch: Base Ten Blocks
Online: Place-Value Dice (hundreds, tens, and ones) and Place-Value Operations Board

GROUPING   Work individually or in pairs.

GETTING STARTED   Begin with the tens and ones dice. One student rolls the dice; the second student uses rods and cubes to construct the two-digit number represented by the dice on the place-value operations board. The number should be constructed first using the least number of blocks possible. One of the highest value blocks should then be traded for 10 of the next lower value blocks. Each number should be recorded as shown in the example. The trading should continue until only unit cubes are used. Students then switch roles.

Example:

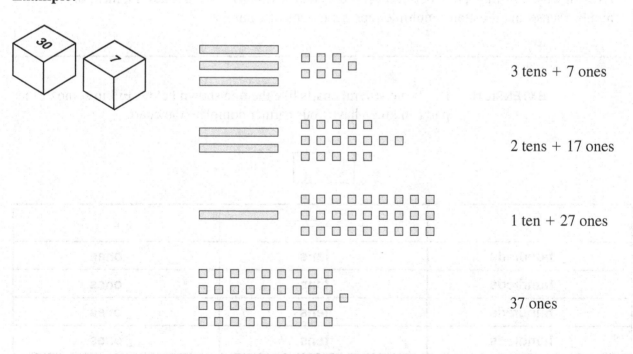

3 tens + 7 ones

2 tens + 17 ones

1 ten + 27 ones

37 ones

Repeat this activity with three place-value dice.

Complete the chart.

243

▨	▬	▫
2 hundreds	4 tens	3 ones
1 hundreds	__ tens	3 ones
1 hundreds	12 tens	__ ones
2 hundreds	3 tens	__ ones
0 hundreds	23 tens	__ ones
__ hundreds	12 tens	23 ones

Explain why rewriting a number in several different ways promotes the development of number sense and illustrates multiple representations of a number.

EXTENSION   Duplicate several charts like the one shown below. Fill in some of the parts and then have your partner complete the chart.

▨	▬	▫
__ hundreds	__ tens	__ ones
__ hundreds	__ tens	__ ones
__ hundreds	__ tens	__ ones
__ hundreds	__ tens	__ ones
__ hundreds	__ tens	__ ones
__ hundreds	__ tens	__ ones

Activity 2: Find the Missing Numbers

PURPOSE   Reinforce number sense and place-value concepts in a problem-
              solving situation.

COMMON CORE SMP   SMP 1, SMP 2, SMP 4, SMP 5, SMP 6

MATERIALS   Pouch: Base Ten Blocks
              Online: Place-Value Operations Board

GROUPING   Work in pairs.

GETTING STARTED   One student reads the clues to the other. The second student uses
              blocks to construct a model of the missing number(s). Students
              alternate turns.

---

1. I have four base ten blocks. Some are
    rods and some are units. Their value is
    less than 25.

    Who am I? _____

---

2. I have eight base ten blocks. Some are
    units and some are rods. Their value is
    an odd number between 50 and 60.

    Who am I? _____

---

3. I have six base ten blocks. Some are
    rods and some are units. Their value is
    between 20 and 40.

    Who am I? _____

---

4. I have six base ten blocks. Some are
    flats, some are rods, and some are
    units. I am a palindrome.

    Who am I? _____

---

5. I have two base ten blocks.

    Who am I? _____

---

6. I have three base ten blocks. None of
    them are flats.

    Who am I? _____

---

7. I have four base ten blocks. None of
    them are flats.

    Who am I? _____

---

8. I have four base ten blocks. Only one
    of them is a flat.

    Who am I? _____

EXTENSION   Create additional clue cards for your partner to solve.

Activity 3:   It All Adds Up

PURPOSE	Develop an understanding of addition of multi-digit numbers and regrouping.
COMMON CORE SMP	SMP 4, SMP 5, SMP 6, SMP 7
MATERIALS	Pouch: Base Ten Blocks
	Online: Place-Value Dice (tens and ones) and Place-Value Operations Board
GROUPING	Work in pairs.
GETTING STARTED	The first student rolls the dice and constructs the two-digit number on a place-value operations board using the correct base ten blocks as in the example. The number should also be recorded on a student record sheet. Then the second student rolls the dice and constructs his or her number beneath that of the first student as shown. The number is also recorded below the first one on the record sheet and a line is drawn across the sheet below the two numbers. The blocks are then moved together to represent the sum of the two numbers. If a column has 10 or more blocks, then a 10-for-1 trade should be made. Record the representation of the sum.

Example: **Place-Value Operations Board**   **Student Record Sheet**

Student 1

Student 2

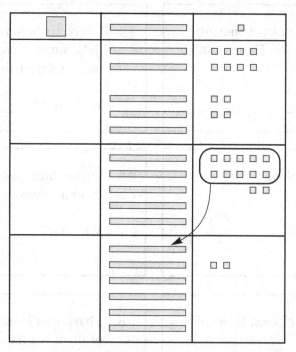

2 tens + 8 ones   28

3 tens + 4 ones   + 34

5 tens + 12 ones

Make 10-for-1 trade

6 tens + 2 ones =   62

1. Roll the dice to make five additional addition problems. Find the sums as shown above.

2. You may be familiar with the use of the term *carry* to describe part of the addition algorithm. Explain how *regrouping* or *renaming* more accurately describes this process than the word *carry*.

Use of base ten blocks in the previous activity promotes the understanding of place value. The use of the blocks also leads to a natural left-to-right progression in the addition of two or more multi-digit numbers. However, once students move from concrete materials to the traditional addition algorithm, the place value inherent in the position of the digits is lost, and the process becomes entirely digit-based.

Example:

$$\overset{1}{3}7$$
$$+\ 28$$
$$\overline{65}$$

The words used to describe this process are: "Seven plus eight is 15; write down the five and carry the one. Two plus three is five plus one is six, to get 65." The place value of the digits is completely ignored.

Consider the algorithms shown below. In each case they

- are based on understanding of place value;
- continue the left-to-right progression, as in reading;
- promote stronger proclivity to do mental math; and
- promote fewer errors.

Left-to-Right

```
  37        437
+ 28      + 328
-----     -----
  50        700
+ 15         50
-----     + 15
  65        765
```

Decomposition

```
  37 →     30 + 7          437 →     400 + 30 + 7
+ 28 →   + 20 + 8        + 328 →   + 300 + 20 + 8
         --------                  ----------------
         50 + 15 = 65              700 + 50 + 15 = 765
```

Making Nice Numbers

$37 + 28 =$

$37 + 30 = 67$

I added on 2 to 28 to make a nice number 30. So now I have to take the 2 away.

$67 - 2 = 65$

$37 + 28 =$

$40 + 28 = 68$

I added on 3 to 37 to make a nice number 40. So now I have to take the 3 away.

$68 - 3 = 65$

Compensation

```
   37(−2) =     35        348(+2) =     350
+  28(+2) =   + 30      + 536(−2) =   + 534
              ----                    -----
                65                      884
```

Use some of the above strategies to solve these problems.

1. $78 + 93 =$ 2. $413 + 289 =$ 3. $57 + 248 =$

4. $544 + 58 =$ 5. $69 + 43 =$ 6. $2006 + 483 =$

Activity 4: What's the Difference?

PURPOSE	Use the *take-away* and *comparison* models to develop an understanding of subtraction of multi-digit numbers.
COMMON CORE SMP	SMP 4, SMP 5, SMP 6, SMP 7
MATERIALS	Pouch: Base Ten Blocks
	Online: Place-Value Operations Board and two sets of Place-Value Dice (tens and ones)
GROUPING	Work in pairs.
GETTING STARTED	For the *take-away* model of subtraction, each student rolls a set of dice. The person with the larger number constructs it on the operations board using base ten blocks. The other student removes (*takes away*) the number of blocks represented by the two-digit number on his or her dice. Record the representation of the difference.

Example 1:

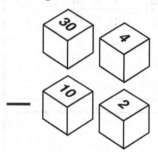

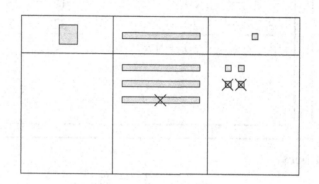

$$\begin{array}{r} 34 \\ -12 \\ \hline 22 \end{array}$$

Example 2:   If necessary, make a 1-for-10 trade.

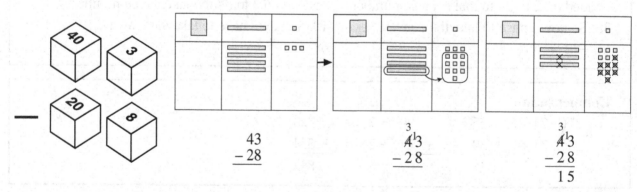

Roll the dice to make five additional subtraction problems. Determine the difference between each pair of numbers using the *take-away* model. Record each subtraction problem and the solution. Show *regrouping* if necessary.

The *comparison* model is a powerful tool to promote the understanding of word problems containing questions such as ... who has more/fewer? ... how much more/less? ... how many more/fewer? These require a comparison of the two amounts and determining how many more/fewer. Model the subtraction shown in Example 3 by constructing the numbers on the dice with base ten blocks. To make the comparison, stack the blocks for the smaller number on top of those for the larger number. If necessary, make a 1-for-10 trade. The blocks not covered represent the difference between the two quantities.

Example 3:

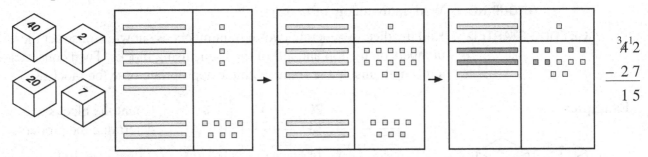

$$\overset{3}{\cancel{4}}{}^{1}2$$
$$-\ 2\ 7$$
$$\overline{1\ 5}$$

---

1. Roll the dice to make five additional subtraction problems. Determine the difference between each pair of numbers using the *comparison* model. Record each subtraction problem and the solution. Show *regrouping* if necessary.

2. Write two word problems for the *take-away* model for subtraction, and two for the *comparison* model. Each problem should be a real world application relevant to the life of an elementary student and correctly represent the designated model.

Alternative algorithms can eliminate the need for regrouping in subtraction.

Counting Up or Counting On

$$\begin{array}{r} 34 \\ -\ 18 \end{array} \quad \begin{array}{l} 18 + 10 = 28 \\ +\ 2 = 30 \\ +\ 4 = 34 \\ \hline \ 16 \end{array}$$

Making Easier but Equal Problems

by Adding		by Subtracting	
44 (+2)	46	3002 (−3)	2999
− 18 (+2)	− 20	− 1267 (−3)	− 1264
	26		1735

Activity 5:   Multi-digit Multiplication

PURPOSE   Develop the multiplication algorithm with multi-digit numbers and reinforce place-value concepts.

COMMON CORE SMP   SMP 4, SMP 5, SMP 6, SMP 7

MATERIALS   Pouch: Base Ten Blocks

Online: Place-Value Dice (tens and ones) and Multiplication and Division Frame

GROUPING   Work individually or in pairs.

GETTING STARTED   Roll the dice to generate two-digit numbers as shown. Construct these numbers on the top and left of the frame using blocks. Then construct a rectangle inside the frame with the appropriate base ten blocks.

Example:

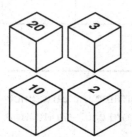

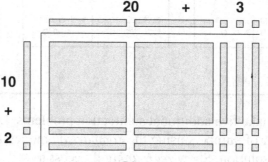

Count the blocks to determine the product.

2 flats   = 200

7 rods   = 70

6 units   =   6

Product = 276

Record your work in one of the expanded notation formats to reinforce place value in the factors and in the partial products.

$$
\begin{array}{rrr}
 & 20 & + & 3 \\
\times & 10 & + & 2 \\
\hline
 & 200 & + & 30 \\
 & & & 40 & + & 6 \\
\hline
 & 200 & + & 70 & + & 6 & = & 276
\end{array}
$$

$$
\begin{array}{r}
23 \\
\times 12 \\
\hline
200 \\
30 \\
40 \\
+\ 6 \\
\hline
276
\end{array}
$$

Roll the place-value dice to generate factors for five additional multiplication problems. Solve the problems using a multiplication and division frame as illustrated in the example above. Record all answers in an expanded form.

EXTENSION   Use a multiplication and division frame and base ten blocks to model multiplication of binomials as shown above. Let a flat equal x^2, a rod equal x, and a unit cube equal 1.

Example: $x + 2 =$ ▭ + ☐ ☐

Complete the following:

1. $(x + 2)(x + 3) =$ 2. $(x + 4)(x + 5) =$

3. $(x + 1)(x + 7) =$ 4. $(x + 6)(x + 8) =$

Activity 6:   Least and Greatest

PURPOSE   Use the guess and check problem-solving strategy to develop number and operation sense and to reinforce the concept of place value and the distributive property.

COMMON CORE SMP   SMP 1, SMP 3, SMP 6, SMP 7, SMP 8

MATERIALS   Other: A calculator

GROUPING   Work individually.

---

1. Arrange the digits 1, 2, 3, 4, and 5 to make a two-digit and a three-digit number. Each digit may be used only once. Use a calculator to multiply the numbers. Try several different arrangements of the digits to determine the arrangement that results in the **greatest** possible product.

    __ __ __   __ __ __   __ __ __
    × __ __   × __ __   × __ __
    _____   _____   _____

2. Repeat the problem and arrange the digits so that you obtain the **least** possible product.

    __ __ __   __ __ __   __ __ __
    × __ __   × __ __   × __ __
    _____   _____   _____

3. Analyze the results in Exercises 1 and 2. Given any five non-zero digits $a < b < c < d < e$, what placement of the digits in a two-digit and a three-digit number will guarantee the **greatest** product and the **least** product? Explain why you placed the digits in each place-value position.

EXTENSIONS   Given any seven different non-zero digits, what placement of the digits in a two-digit number and a five-digit number, or a four-digit and a three-digit number, will guarantee the **greatest** product and the **least** product?

Does the arrangement of five or seven digits also work for arranging six digits in two three-digit numbers to guarantee the **greatest** product? Explain.

Activity 7: Multi-digit Division

PURPOSE	Develop the division algorithm and reinforce the *partitioning model* and the place-value concepts associated with division.
COMMON CORE SMP	SMP 4, SMP 5, SMP 6, SMP 7
MATERIALS	Pouch: Base Ten Blocks
	Online: Place-Value Dice (hundreds, tens, and ones) and Place-Value Operations Board
	Other: Paper squares approximately 15 cm square
GROUPING	Work individually or in pairs.
GETTING STARTED	Roll the place-value dice to generate a three-digit product (the dividend) and use base ten blocks to construct it on a place-value operations board. Roll the ones die to determine the factor (divisor). The divisor determines the number of paper squares needed to represent the number of groups.

Example: $5\overline{)167}$

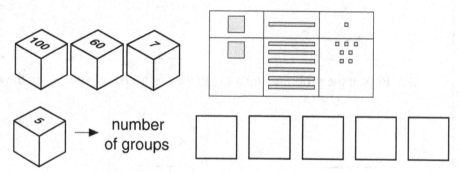

Solution:

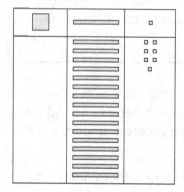

a.   Place an equal number of the highest value blocks (flats) on each square. If this is not possible, record a 0 in the hundreds place above the division frame as shown below. Make a 1-for-10 trade with the next lower value block (rods) as shown at the left. In some cases, it may be necessary to make more than one trade.

$$\frac{0}{5\overline{)167}}$$

b.   There are now 16 rods. Place an equal number of rods on each square. What is the value of the rods on each square? _____

Record these results as follows:

$$
\begin{array}{r}
030 \\
5\overline{)167} \\
\text{tens} \\
\underline{150} \\
17
\end{array}
$$

c. Proceed as shown on the previous page. Make a 1-for-10 trade and share an equal number of unit blocks among the squares. Record the results of the sharing as shown.

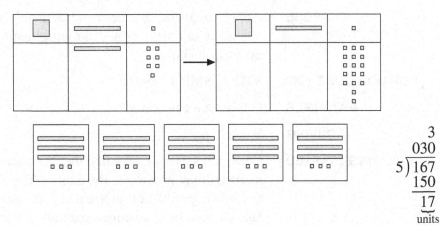

$$\begin{array}{r} 3 \\ 030 \\ 5\overline{)167} \\ 150 \\ \hline 17 \\ \underset{\text{units}}{\smile} \\ 15 \end{array}$$

What is the value of the unit blocks on each square? _____

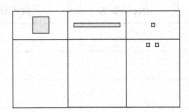

How many blocks remain? _____ Can they be shared equally among the squares? _____

Record the final results as follows:

When 167 is distributed among 5 groups, there are 33 in each group and 2 left over that cannot be shared (the remainder).

$$\begin{array}{r} 3 \\ 030 \\ 5\overline{)167} \\ 150 \\ \hline 17 \\ 15 \\ \hline 2 \end{array} = 33 \text{ remainder } 2$$

Use the place-value dice to generate a factor (divisor) and a product (dividend) for six problems. Use base ten blocks and the method described above to model the division. Record results as illustrated in the example.

EXTENSION   Explain the role that *regrouping* (*renaming*) of numbers plays in multiplication and division with multi-digit numbers.

Activity 8: Target Number

PURPOSE	Apply the guess and check problem-solving strategy to reinforce the inverse relationship between multiplication and division, and develop estimation skills.
COMMON CORE SMP	SMP 1, SMP 4, SMP 5
MATERIALS	Other: One calculator
GROUPING	Work in pairs.
GETTING STARTED	Construct several tables like the one shown below. Only whole numbers may be used in the games. The calculator may only be used to find the product or quotient as indicated in each game. All other factors must be determined mentally.

TARGET = _____			
Turn	**Factors**	**Product**	**Difference**
1			
2			
3			
		TOTAL	

GAME 1

- Players choose a target number, and each player records it in a table.

- On alternate turns, each player chooses two factors (other than 1 and the target) and records them in the table. Then the player uses the calculator to multiply the factors and records the product.

- If the product equals the target, the player wins. If not, the player records the absolute value of the difference between the product and the target number in the table.

- If neither player has won after three turns, the person with the least total for the three differences is the winner.

GAME 2

- Player 1 chooses a target number; Player 2 chooses a constant factor and an acceptable range for the answer, for example ±20.
- On alternate turns, each player chooses another factor and uses the calculator to multiply it by the constant factor to obtain a product.
- If the absolute value of the difference between the product and the target number is within the range chosen by the players, the player wins.
- If the absolute value of the difference is outside the range, the players alternate turns until one player obtains a product within the specified range of the target number.

SAMPLE GAME

Target Product = 650   Constant Factor = 19   Range = ±8

	Constant Factor		Estimated Factor		Product	| Difference |
Player 1	19	×	37	=	703	53
Player 2	19	×	30	=	570	80
Player 1	19	×	33	=	627	23
Player 2	19	×	34	=	646	4

Player 2 wins!

EXTENSIONS

1. Repeat **GAME 1** using division. On each turn, the player chooses a dividend and a divisor and divides them. If the quotient equals the target number, the player wins. If not, the player records the difference between the quotient (rounded to the nearest whole number) and the target in the table.

2. Repeat **GAME 2** using division. Player 1 chooses the target quotient. Player 2 chooses the constant divisor and an acceptable range for the answer.

SAMPLE GAME 2

Target Quotient = 26   Constant Divisor = 14   Range = ±1

	Estimated Dividend		Constant Divisor		Quotient	| Difference |
Player 1	320	÷	14	≈	22.9	3.1
Player 2	380	÷	14	≈	27.1	1.1
Player 1	365	÷	14	≈	26.07	0.07

Player 1 wins!

Activity 9:   A Visit to Fouria

PURPOSE   Use a game to reinforce place-value concepts, to introduce the base-four numeration system, and to develop understanding of the regrouping process.

COMMON CORE SMP   SMP 1, SMP 2, SMP 4, SMP 5, SMP 6

MATERIALS   Other: Blue, red, and white chips (10 of each color) and a die

GROUPING   Work in pairs.

GETTING STARTED   While on an Intergalactic Numismatics Tour you encounter a meteor shower and are forced to make an unscheduled stop on the planet Fouria. The monetary system used on Fouria consists of three coins: a white coin (worth $1 in U.S. money), a red coin, and a blue coin. The red coin is equivalent in value to four white coins, and the blue coin is equivalent to four red coins.

1. Unlike its sister planet, Ufouria, Fouria turns out to be a rather dull place to visit. To help pass the time in the waiting area, you and a fellow passenger play a coin trading game. The rules of the game are:

    • Players alternate turns.

    • On each turn, a player rolls the die and places that number of white Fourian coins in the white column on his or her Coin Trading Game Sheet.

    • Whenever possible, a player must trade four white coins for one red coin and/or four red coins for one blue coin.

    • Coins must be placed in the appropriately labeled column. No more than three coins of one color may be in any column at the end of a turn.

    • The first player to get two blue coins is the winner.

    Make a Coin Trading Game Sheet, and play the game with a partner. At the end of each turn, record the number of each color coin on your game sheet in a table like the one at the right.

Coin Trading Game Sheet		
Blue	**Red**	**White**

Turn	Number Rolled	Result		
		B	**R**	**W**
1				
2				
3				
4				
5				
6				
7				
8				
9				
10				

THE COIN TRADING GAME REVISITED

1. A third passenger has been watching you play. She suggests it is more challenging to start the game with three blue coins and to remove the number of white coins equal to the number rolled on each turn. The first player to remove all the coins from his or her playing board is the winner. Your playing partner is confused. "How can you remove white coins when there aren't any on the board?" he asks. Explain how this can be done.

2. Play this new version of the game with a partner. Each player should record the result of each move in a table like the one on the preceding page.

---

1. You become bored with the games and go to the spaceport newsstand to buy something to read. Glancing at the cover of a Fourian magazine you notice that the price is given as 123_{four}. At the checkout stand, the clerk explains that this means one blue coin, two red coins, and three white coins. How many of each color coin does each of the following prices represent?

   a. 231_{four} _____

   b. 102_{four} _____

   c. 13_{four} _____

   d. 20_{four} _____

2. How would Fourians write each of the following prices?

   a. 1 blue, 2 red, 1 white _____

   b. 2 red, 3 white _____

   c. 2 blue _____

   d. 2 blue, 3 white _____

3. Back in the waiting area, you begin leafing through a magazine that you purchased. You note that the first page is numbered 1_{four}, but when you get to the fourth page, you are surprised to find that it is numbered 10_{four}. Fill in the blanks below to show how the remaining pages of the magazine would be numbered.

   1_{four} _____ _____ 10_{four} _____ _____ _____ _____ _____

   _____ _____ _____ _____ 33_{four} _____ _____ _____ _____ _____

   _____ _____ _____ _____ _____ _____ _____ · · ·

   _____ _____ 333_{four} _____ _____ _____ _____ _____ _____

1. After reading for a while, you decide to have dinner. The price of your meal was 121_{four}. When you go to the cashier to pay for your meal, you realize that you don't have any Fourian money with you. "No problem," the cashier explains. "You may pay with dollars." What is the cost of your meal in dollars?

2. a. On your way back to the waiting area, you stop in the newsstand to buy a souvenir. It costs 1312_{four}. How many dollars is this?

   b. You give the cashier two $100 bills. How much Fourian money should you get back in change?

---

1. Back in the waiting area, you find a mathematics book left behind by a Fourian student. Flipping through the book, you come across the examples shown below. Explain what the small 1 in the second example means and how it was obtained.

$$\begin{array}{r} 2\,3\,1_{four} \\ +\,1\,0\,2_{four} \\ \hline 3\,3\,3_{four} \end{array} \qquad \begin{array}{r} 2\overset{1}{}3_{four} \\ +\ \ 3\,3_{four} \\ \hline 1\,2\,2_{four} \end{array}$$

2. Use what you learned in the examples in Exercise 1 to find the following sums.

   a. $121_{four} + 211_{four}$ b. $123_{four} + 221_{four}$

3. A few pages later, you find the following examples. Explain what is being done in steps A, B, and C.

$$\begin{array}{r} 3\,3\,3_{four} \\ -\,1\,2\,1_{four} \\ \hline 2\,1\,2_{four} \end{array}$$

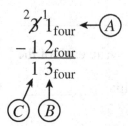

4. Use what you learned in the examples in Exercise 3 to find the following differences.

   a. $323_{four} - 211_{four}$ b. $221_{four} - 132_{four}$

EXTENSIONS

1. Explain how the place-value system used to write Fourian numerals is related to the place-value system used to write base-ten numerals.

2. Explain how the regrouping used in Fourian addition and subtraction is related to the regrouping used in traditional (base-ten) addition and subtraction.

Chapter Summary

"Number sense refers to an intuitive feeling for numbers and their uses and interpretations; an appreciation for various levels of accuracy when figuring; the ability to determine arithmetical errors; and a common sense approach to using numbers."

—*Developing Number Sense*
Addenda Series, Grades 5–8

The development of number sense, operation sense, and computational fluency is a focal point of the mathematics curriculum described in the *Principles and Standards for School Mathematics* and the *Curriculum Focal Points for Prekindergarten through Grade 8 Mathematics*.

Many of the activities in this chapter illustrate a curriculum design that begins with concrete models such as base ten blocks, and proceeds through representations (connecting models and symbolism) to the abstract (numeral).

Activities 1 and 2 developed number sense and an understanding of place-value and regrouping concepts in our base-ten numeration system.

Activities 3 and 4 developed the algorithms for addition and subtraction of whole numbers that were modeled on the Place-Value Operations Board. Recording each step of the activity on the Student Record Sheet helped develop a clear understanding of the addition and subtraction algorithms as well as the associated place-value concepts.

The *comparison* model for subtraction was emphasized to promote understanding of problems containing phrases such as "how many more (greater) than?" and "how many less (fewer) than?" This model also illustrated the concept of subtraction as the inverse of addition. When the blocks representing the number to be subtracted were placed on top of the blocks of the larger number (see Example 3 in Activity 4), the remaining blocks represented the number that must be *added* to the smaller number to obtain the larger.

Activities in Chapter 2 developed a basic understanding of multiplication and division and also illustrated the concept of inverse operations. In Activities 5 through 7, this basic understanding was extended to the use of the algorithms with multi-digit numbers. The two words *factor* and *product* were stressed in both of these operations rather than the traditional *multiplier*, *multiplicand*, *dividend*, *divisor*, and *quotient*. This focus helps to reinforce the inverse relationship between the operations.

Activities 3 through 5 also included alternative algorithms for addition, subtraction, and multiplication. These algorithms are based on number sense and are consistent with the left-to-right process illustrated in the base-ten blocks models for the algorithms. The alternative methods are *place-value* based rather than *digit* based. They require less instructional time by teachers, and less memorization of rote procedures by students. They also are much less prone to procedural errors and are consistent with the left-to- right process used in reading and students' early mathematics work with patterns, number lines, and hundreds charts.

In Activities 6 and 8, a calculator was used in problem-solving settings to reinforce concepts in earlier activities. Investigation of the partial products obtained from multiplying the various arrangements of the digits in Activity 6 develops a deeper understanding of place value and leads to the arrangement of digits needed to obtain the greatest or least product. Estimation and mental arithmetic strategies are important features of the Target Number activity.

Activity 9 reinforced the understanding of our base ten numeration system through the investigation of a base-four numeration system.

The relationships among the four basic operations are illustrated in two different diagrams below. Notice the connections between multiplication as repeated addition and division as repeated subtraction.

	Once	Many
Join	+	×
Separate	−	÷

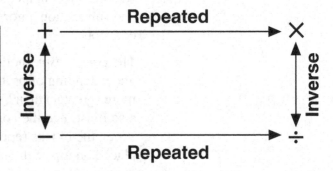

Chapter 4
Number Theory

"Throughout the study of numbers, students … should identify classes of numbers and examine their properties. For example, integers that are divisible by 2 are called *even numbers* and numbers that are produced by multiplying a number by itself are called *square numbers*. Students should realize that different types of numbers have particular characteristics; for example, square numbers have an odd number of factors and prime numbers have only two factors."
> —*Principles and Standards for School Mathematics*

"Number theory offers many rich opportunities for explorations that are interesting, enjoyable, and useful. These explorations have payoffs in problem solving, in understanding and developing other mathematical concepts, in illustrating the beauty of mathematics and in understanding the human aspects of the historical development of number."
> —*Curriculum and Evaluation Standards for School Mathematics*

Number theory is primarily concerned with the study of the properties of the natural numbers. Number theory topics, such as multiples, factors, prime numbers, prime factorization, least common multiples, and greatest common divisors, are an integral part of the elementary and middle school curricula. These concepts are used extensively when working with rational numbers.

In this chapter, you will use a variety of formats—concrete models, games, a grid-coloring activity, and geometric applications—to explore topics in number theory. The concrete models, coloring activity, and the geometric applications develop visual images of the number theory concepts involved. These multiple representations promote greater conceptual understanding by connecting the abstract ideas and the physical models that are used.

Correlation of Chapter 4 Activities to the
Common Core Standards of Mathematical Practice

Activity Number and Title	Standards of Mathematical Practice
1: A Square Experiment	SMP 3, SMP 4, SMP 5, SMP 7
2: The Factor Game	SMP 2, SMP 4, SMP 5, SMP 7, SMP 8
3: A Sieve of Another Sort	SMP 2, SMP 3, SMP 4, SMP 5, SMP 7, SMP 8
4: Great Divide Game	SMP 5, SMP 6, SMP 7
5: Tiling with Squares	SMP 2, SMP 3, SMP 4
6: Pool Factors	SMP 4, SMP 5, SMP 6, SMP 7

Activity 1: A Square Experiment

PURPOSE	Develop the concepts of prime, composite, and square numbers using a geometric model.
COMMON CORE SMP	SMP 3, SMP 4, SMP 5, SMP 7
MATERIALS	Pouch: Colored Squares Online: Centimeter Graph Paper
GROUPING	Work individually or in groups of 3 or 4.
GETTING STARTED	Use squares or graph paper to form all the rectangular arrays possible with each different number of squares. Record your results in Table 1.

Examples: Number of Squares   1   2   3

Rectangular Arrays

Note: ⬜⬜ is not a rectangular array.

When an array is described by its dimensions, the figure ⬜⬜ has an altitude of 1 unit and a base of 2 units and is labeled 1×2. The figure ⬛ is labeled 2×1.

TABLE 1

Number of Squares	Dimensions of the Rectangular Arrays	Total Number of Arrays
1		
2		
3	$1 \times 3, 3 \times 1$	2
4		
5		
6	$1 \times 6, 6 \times 1, 3 \times 2, 2 \times 3$	4
7		
8		
9		
10		
11		
12		

1. Use the results from Table 1 to complete Table 2.

TABLE 2

Number of Squares That Produced:			
A	**B**	**C**	**D**
Only One Array	**Only Two Arrays**	**More Than Two Arrays**	**An Odd Number of Arrays**

2. Suppose you have 24 squares.

    a. How many rectangular arrays can be made?

    b. In which column(s) in Table 2 would you place 24?

    c. What are the factors of 24?

3. a. What are the factors of 16?

    b. How many rectangular arrays can you make with 16 squares?

    c. In which column(s) of Table 2 would you place 16?

4. Look at the data in Table 1 and Table 2. How is the number of factors of a given number related to the number of rectangular arrays?

5. a. Why is it that the numbers in column D of Table 2 produce an odd number of arrays?

    b. What are the next three numbers that would be placed in column D?

6. What is the mathematical name for the numbers in

    a. column B?

    b. column C?

    c. column D?

7. Which numbers can be placed in two lists? Why?

8. Can any numbers be placed in three lists? If so, which ones?

9. Write each of the following composite numbers as a product of primes.

    a. 28   b. 42

    c. 150   d. 231

10. a. Can every composite number be written as a product of primes? Explain your reasoning.

    b. If two people write the same number as a product of primes,

        i. how would their factorizations be alike?

        ii. how might the factorizations be different?

Activity 2: The Factor Game

PURPOSE	Develop the idea of the prime factorization of a number using a game.
COMMON CORE SMP	SMP 2, SMP 4, SMP 5, SMP 7, SMP 8
MATERIALS	Pouch: 15 each, two different Colored Squares Other: Six paper clips and the Factor Game Game Board (page 63)
GROUPING	Work in pairs or in teams of two students each.
GETTING STARTED	Play the *Factor Game* three times with your partner.

- To begin, one player places two paper clips on numbers in the factor list. The paper clips may be placed on the same number or on different numbers. The player then multiplies the numbers and places a square of his or her color on the product on the game board.

- Players then alternate turns. On a turn, a player may form a new product in **one** of the following ways:

  A. Place a new paper clip on any number in the factor list.

  B. Take one paper clip off a number in the factor list.

  C. Move one of the paper clips already on the factor list to a different number.

- A turn ends when a player places a colored square on a product, or the product is already covered or it is not on the game board.

- The first player to cover three adjacent products in a row horizontally, vertically, or diagonally with squares of his or her color is the winner.

When you have finished playing the game, answer the following questions.

1. Are the numbers on the game board *prime* or *composite*? the numbers in the factor list?

2. a. List all of the factors of 60. Identify which factors are prime, which are composite, and which are neither prime nor composite.

    b. List all the different ways that paper clips can be placed on numbers in the factor list to result in a product of 60.

Factor Game

Game Board

4	12	25	90	36
108	75	120	6	10
15	80	8	9	45
54	30	100	18	50
20	24	27	40	60

Factor List

2	3	5

Every composite number can be written as the product of prime factors. Such a product is the **prime factorization** of the number.

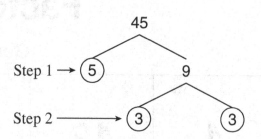

A *factor tree* can be useful for finding the prime factorization of a number. An example of a factor tree for 45 is given at the right.

1. Describe what happens in Step 1 of the factor tree for 45. Why is the 5 circled but not the 9?

2. Describe what happens in Step 2. Why do the branches stop at the circled numbers?

3. What is the prime factorization of 45?

---

1. Complete the factor tree for 120 at the right. Why are more branches necessary to make this tree than to make the one for 45?

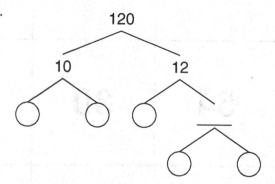

2. What is the prime factorization of 120?

3. Sketch a factor tree of your own for 120 that starts with a different pair of factors.

4. Compare the circled numbers at the ends of the branches in the two factor trees for 120. What do you notice?

Activity 3: A Sieve of Another Sort

PURPOSE   Investigate primes, composites, multiples, and prime factorizations using a "sieve."

COMMON CORE SMP   SMP 2, SMP 3, SMP 4, SMP 5, SMP 7, SMP 8

MATERIALS   Other: Orange, red, blue, green, and yellow colored pencils or crayons

GROUPING   Work individually.

GETTING STARTED   Eratosthenes, a Greek mathematician, invented the "sieve" method for finding primes over 2200 years ago. This activity explores a variation of Eratosthenes' sieve.

As you discovered in Activity 1, *one* is neither prime nor composite. To show this, mark an X through 1.

The first prime number is 2. Color the diamond in which 2 is located **orange**. Use **red** to color the upper-left corner of the **Key** and the upper-left corner of all squares containing multiples of 2. Any number greater than 2 with a corner colored will fall through the sieve.

What was the first multiple of 2 that fell through the sieve? _____

The next uncolored number is 3. Color the diamond surrounding the 3 **orange**. Use **blue** to color the upper-right corner of the **Key** and the upper-right corner of all squares containing multiples of 3.

What was the first multiple of 3 that fell through the sieve? _____

Repeat this process for 5 and 7. Color the diamonds surrounding the numbers **orange**. Use **green** to color the lower-right corners of the **Key** and of all squares containing multiples of 5. Use **yellow** to color the lower-left corners of the **Key** and of all the squares containing multiples of 7. Note the first multiples of 5 and of 7 that fall through the sieve.

Finally, use **orange** to color the diamond surrounding all the numbers in the grid that are in squares with no corners colored. These numbers are all primes.

---

1. How do you know that 2, 3, 5, and 7 are prime numbers?

2. How can you tell that 2, 3, 5, and 7 are prime numbers from the way the sieve is colored?

3. How can you identify composite numbers from the way the sieve is colored?

When you colored the multiples of 2, the number 4 was the first multiple of 2 that fell through the sieve.

When you colored the multiples of 3, the number 9 was the first multiple of 3 that fell through the sieve.

1. When you colored multiples, what was

    a. the first multiple of 5 that fell through the sieve?

    b. the first multiple of 7 that fell through the sieve?

After you colored the multiples of 7, the next uncolored number was 11.

2. If you could color multiples of 11, what would be the first number to fall through the sieve?

3. When you color multiples of a prime number, how is the first multiple of the prime that falls through the sieve related to the prime number?

4. If the grid went to 300, what is the largest prime whose multiples must be colored before you can be certain that all of the remaining uncolored numbers are prime?

5. What is the largest prime less than 1000? Explain how you obtained your answer.

The sieve can be used for more than finding primes.

1. List the numbers that are colored with the code for 2 and for 3.

2. The numbers in Exercise 1 are multiples of both 2 and 3. What numbers are they?

3. How could you use the color code to find

   a. the multiples of 14?

   b. the multiples of 30?

The sieve can also help you find the prime factorization of a number.

Example: From the sieve, you find that the $72 = 2 \times 3 \times 12 \leftarrow 12$ is not prime
prime factors of 72 are 2 and 3.

$$2 \text{ is prime}$$
$$\downarrow$$

Again from the sieve, the prime $72 = 2 \times 3 \times (2 \times 3 \times 2)$
factors of 12 are 2 and 3. $= 2 \times 2 \times 2 \times (3 \times 3)$

Since 2 is a prime, you are done. $72 = 2^3 \times 3^2$

1. Find the prime factorization of

   a. 54   b. 84   c. 100

---

1. Pairs of prime numbers like 3 and 5 that differ by 2 are called *twin primes*. List all the twin primes less than 100.

2. What is the longest string of consecutive composite numbers on the grid?

3. Several of the numbers on the grid are divisible by three different primes. What is the smallest number that is divisible by four different primes?

Activity 4: Great Divide Game

PURPOSE	Apply and reinforce divisibility rules.
COMMON CORE SMP	SMP 5, SMP 6, SMP 7
MATERIALS	Other: Great Divide Game Board (page 69), a number cube, and chips for markers
GROUPING	Work in groups of two to four players.
GETTING STARTED	Follow the rules below to play the *Great Divide Game*.

RULES FOR THE GREAT DIVIDE GAME

1. Make a number cube labeled as follows.   2, 3, 4, 5, 5, 9

2. To begin the game, each player rolls the number cube. The player with the greatest number goes first; play progresses to the next player on the right.

3. On each turn, the player rolls the number cube and places a chip on a number on the game board that is divisible by the number showing on the number cube.

    Example: Player rolls 3   Player covers 168   Score 1 point

    If the player can name other numbers on the number cube that are factors of the number that was covered, the player scores one point for each.

           Player names 2 and 4   Score 2 points

                     Total: 3 points for that turn

4. If the player is unable to find an uncovered number on the game board that is divisible by the number showing on the number cube, he or she must pass the number cube to the next player. If another player knows a play that can be made with the number on the number cube, that player may call attention to the mistake and tell the other players what uncovered number on the board is divisible by the number on the cube. The player citing the mistake may then place a chip on that number and earn points. This does not affect the turn of the player citing the mistake. If more than one player calls attention to a mistake, the first player to do so makes the play.

5. Players keep a running total of their scores. A player who cannot cover a number in three successive turns is eliminated from the game. When the game board is filled, or if all players have failed to play in three successive turns, the game ends. The player with the highest score is the winner.

Great Divide

Game Board

168	435	105	702	180	156
444	342	270	189	387	258
231	390	459	260	324	465
280	153	388	396	477	294
152	138	110	594	154	315
279	666	515	340	195	378

Divisibility Rules

2 A number is divisible by 2 if it is an *even* number.

3 A number is divisible by 3 if the sum of the digits is divisible by 3.

4 A number is divisible by 4 if the two-digit number formed by the tens and ones digits is divisible by 4.

5 A number is divisible by 5 if the ones digit is 0 or 5.

9 A number is divisible by 9 if the sum of the digits is divisible by 9.

Activity 5: Tiling with Squares

PURPOSE   Use a geometric model to explore the concept of the greatest common divisor (GCD) of two numbers and use the model to develop an algorithm for finding the GCD.

COMMON CORE SMP   SMP 2, SMP 3, SMP 4

MATERIALS   Online: Half-centimeter Graph Paper

GROUPING   Work individually or in pairs.

GETTING STARTED   Tiling a region means to completely cover it with non-overlapping shapes. The study of tilings can lead to some interesting questions. For example:

If m and n are whole numbers, what is the length of a side of the largest square that can be used to tile an m × n rectangle?

To answer this question, let's look at some rectangles. For each problem, draw the rectangle on graph paper and either draw or cut out squares to tile it.

1. In Parts a–d, if a 4 × 8 rectangle can be tiled with squares of the given size, make a sketch to show the tiling. If it can't, explain why not.

    a. 1 × 1   b. 2 × 2   c. 3 × 3   d. 4 × 4

2. Can any other size square be used to tile a 4 × 8 rectangle? Explain.

3. The results for a 4 × 8 rectangle are summarized in the table. Fill in the data for the other rectangles listed in the table.

Dimensions of Rectangle	Prime Factorization		Common Prime Factors	Dimensions of Squares That May Be Used	Length of Side of Largest Square
width × length	width	length			
4 × 8	2 × 2	2 × 2 × 2	2 × 2	1 × 1, 2 × 2, 4 × 4	4
6 × 9					
18 × 30					
24 × 36					
8 × 15					

4. How does the length of a side of the largest square that can be used to tile a rectangle appear to be related to the common prime factors of the width and length of the rectangle?

1. a. List the divisors (factors) of 18.

    b. List the divisors of 30.

    c. List the common divisors of 18 and 30.

    d. What is the GCD of 18 and 30?

2. a. List the divisors of 24.

    b. List the divisors of 36.

    c. List the common divisors of 24 and 36.

    d. What is the GCD of 24 and 36?

3. How do the greatest common divisors found in Exercises 1 and 2 compare to the **Length of the Side of the Largest Square** that can be used to tile the corresponding rectangle in the table?

4. If *m* and *n* are whole numbers, what is the length of a side of the largest square that can be used to tile an $m \times n$ rectangle?

---

1. a. List the divisors of 54, 72, and 90.

    b. List the common divisors of 54, 72, and 90.

    c. What is the GCD of 54, 72, and 90?

2. a. Find the prime factorizations of 54, 72, and 90.

    b. List the common prime factors of 54, 72, and 90.

    c. How does the GCD of 54, 72, and 90 appear to be related to the common prime factors of the numbers?

3. a. List the divisors of 84, 105, 126, and 210.

    b. List the common divisors of 84, 105, 126, and 210.

    c. What is the GCD of 84, 105, 126, and 210?

4. a. Find the prime factorizations of 84, 105, 126, and 210.

    b. List the common prime factors of 84, 105, 126, and 210.

    c. How is the GCD of 84, 105, 126, and 210 related to the common prime factors of the numbers?

5. Explain how the prime factorizations of a set of numbers can be used to find the greatest common divisor of the numbers.

Activity 6: Pool Factors

PURPOSE   Apply the concepts of greatest common divisor, least common multiple, and relatively prime numbers in a geometric problem situation.

COMMON CORE SMP   SMP 4, SMP 5, SMP 6, SMP 7

MATERIALS   Online: Centimeter Graph Paper
Other: Straightedge
Alternative: Paper Pool applet on the NCTM Illuminations website:
http://illuminations.nctm.org/ActivityDetail.aspx?ID=28

GROUPING   Work individually or in small groups.

GETTING STARTED   On a piece of graph paper, draw several pool tables like the one shown below, but with different dimensions. Label the pockets *A, B, C,* and *D* in order, starting with the lower-left pocket as shown.

Place a ball on the dot in front of pocket *A.* Shoot the ball as indicated by the arrows. The ball always travels on the diagonals of the grid and rebounds at an angle of 45 degrees when it hits a cushion.

Count the number of squares through which the ball travels.

Count the number of *hits,* that is, the number of times the ball hits a cushion, the initial hit at the dot, and the hit as the ball goes into a pocket.

In the table, enter the dimensions of each pool table, the number of squares through which the ball travels, and the number of hits. Analyze the data in the table and determine a rule that predicts the number of squares and the number of hits, given the dimensions of any pool table.

Height	Base	Number of Hits	Number of Squares
4	6		
5	7		
3	2		

Pool Table

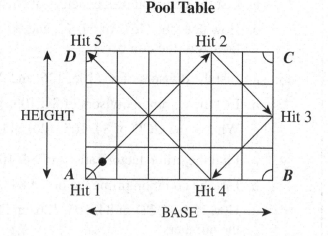

EXTENSIONS   Add a column headed **Final Pocket** to the table. In this column, for each pool table, record the letter of the pocket into which the ball finally fell. Use this data to find a rule that will predict which pocket the ball will fall into for any pool table.

Chapter Summary

In Activities 1 and 3, you discovered that natural numbers can be classified by how many factors they have. *Prime numbers* have exactly two factors; *composite numbers* have more than two; *square numbers* have an odd number of factors; and *one*, which is in a class of its own, has exactly one factor.

Initially, this classification of the natural numbers may have seemed rather arbitrary. However, in Activities 2 and 3, you discovered that every natural number greater than one is either a prime number or can be expressed as a product of prime numbers, and that this product is unique except for the order of the factors. Thus, the prime numbers are the building blocks from which all natural numbers greater than one are constructed. This result is known as the *Fundamental Theorem of Arithmetic*.

The *Great Divide Game* in Activity 4 provided an opportunity for you to reinforce your skills in applying divisibility rules.

Activities 5 and 6 were devoted to the study of factors and multiples of numbers. The *greatest common divisor* (GCD), also known as the *greatest common factor* (GCF), and procedures for calculating it were developed using a geometric model. *Least common multiples* (LCM) and greatest common divisors were applied in Activity 6 and will be used extensively in your study of the rational numbers.

Chapter 5
Integers

"In the lower grades, students may have connected negative numbers in appropriate ways to informal knowledge derived from everyday experiences, such as below-zero winter temperatures or lost yards on football plays. In the middle grades, students should extend these initial understandings of integers. Positive and negative integers should be seen as useful for noting relative changes or values. Students can also appreciate the utility of negative integers when they work with equations whose solution requires them, such as $2x + 7 = 1$."

—*Principles and Standards for School Mathematics*

"By applying properties of arithmetic and considering negative numbers in everyday contexts (e.g., situations of owing money or measuring elevations above and below sea level), students explain why the rules for adding, subtracting, multiplying, and dividing with negative numbers make sense."

—*Curriculum Focal Points for Prekindergarten through Grade 8 Mathematics*

Many everyday situations cannot be adequately described without using both positive and negative numbers. Profit and loss, temperatures above and below 0°, elevations above and below sea level, and deposits and withdrawals are just a few examples. This chapter introduces negative numbers by extending your knowledge of whole numbers to the set of integers.

In Activity 1, ● represents a proton, a subatomic particle with a positive electrical charge of one unit, and ○ represents an electron, a particle with a negative electrical charge of one unit. Because protons and electrons have opposite charges, when a proton and an electron are paired together, they neutralize each other; that is, the pair has an electrical charge of zero. You will use concrete models for integers, like charged-particles, and your understanding of the operations with whole numbers to develop the concept of absolute value and the algorithms for the operations with integers.

Correlation of Chapter 5 Activities to the
Common Core Standards of Mathematical Practice

Activity Number and Title	Standards of Mathematical Practice
1: Charged Particles	SMP 4
2: Coin Counters	SMP 2, SMP 3, SMP 4, SMP 5
3: Subtraction Patterns	SMP 1, SMP 2, SMP 7
4: A Clown on a Tightrope	SMP 2, SMP 4, SMP 5
5: Multiplication and Division Patterns	SMP 1, SMP 2, SMP 7
6: Integer × and ÷ Contig	SMP 2, SMP 7

Activity 1: Charged Particles

PURPOSE	Use the charged-particle model to represent integers and to explore absolute value.
COMMON CORE SMP	SMP 4
MATERIALS	Other: Two different-colored chips (or squares cut from tag board), 15 of each color
GROUPING	Work individually or in small groups.
GETTING STARTED	Use the dark chips to represent protons and the light chips to represent electrons. Construct two different models that represent each integer and sketch your models in the boxes.

Examples:

The set at the right shows one way to represent the number 2.

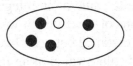

If the protons and electrons are paired, 2 protons are left over. The net electrical charge is 2.

The set at the right shows one way to represent the number −3.

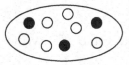

If the protons and electrons are paired, 3 electrons are left over. The net electrical charge is −3.

5	5

−1	−1

−2	−2

0	0

1. Look back at your models on the previous page. If you have not already done so, represent each integer using the fewest number of protons or electrons possible and sketch the model in the space provided.

    a. +5   b. −1   c. −2   d. 0

2. What is the fewest number of particles needed to model each integer in Exercise 1?

    a. _____   b. _____   c. _____   d. _____

The answers to Exercise 2 are the absolute values of the integers in Exercise 1. Since the absolute value of an integer is represented by the fewest number of protons or electrons, it will always be 0 or a positive number. Since the distance between two points is always a positive number or 0, the absolute value of an integer may also be defined as the distance from 0 to the point corresponding to the integer on a number line.

Examples: The absolute value of 7, written as $|7|$, is 7.
$|-8| = 8$.

3. What is the absolute value of each of the following integers?

    a. −15 _____   b. 12 _____   c. 0 _____   d. −5 _____

4. Use your results from the preceding exercises to complete each statement.

    a. The absolute value of a positive integer is _____

    b. The absolute value of a negative integer is _____

    c. The absolute value of 0 is _____

Activity 2: Coin Counters

PURPOSE	Use a game to discover algorithms for integer addition.
COMMON CORE SMP	SMP 2, SMP 3, SMP 4, SMP 5
MATERIALS	Other: A paper cup, 10 pennies, and a game marker
GROUPING	Work in pairs or in groups of 2 or 3.
GETTING STARTED	• At the beginning of the game, each player places a game marker on zero on a number line like the one below.
	• Players alternate turns.
	• On your turn, place six pennies in the cup, cover the opening with your hand, shake the cup thoroughly, and drop the coins onto the table. Each HEAD means move your marker to the right one unit; each TAIL means move it to the left one unit.
	• The first player to go past 10 or −10 is the winner. If there is no winner after ten turns, the player closest to 10 or −10 wins.

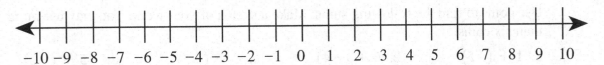

Play the game twice. When you have finished, each player should answer the following questions.

1. Did you find a way to quickly determine where to place your marker after a coin toss? Explain.

2. If you were to represent the number of HEADS with an integer, would you use a positive or a negative integer?

3. Would you use a positive or a negative integer to represent the number of TAILS?

4. Did your marker ever end up an odd number of units away from where it was at the start of your turn? Explain.

5. At the end of a turn, did your marker ever end up in the same place where it started? Explain.

6. Use coins to construct two different representations for each integer. You may use more or less than six coins in a model.

   a. 4   b. −3   c. 0

You have seen how coins can be used to represent integers. Coins can also be used to model addition of integers. Think of the HEADS as a positive integer and the TAILS as a negative integer. For example, tossing 2 HEADS and 4 TAILS is the same as adding 2 and −4.

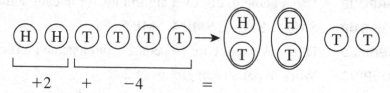

1. a. Why do the paired coins cancel each other out?

    b. If you tossed this combination of coins, how would you move your marker?

    c. What integer is represented by the combination of coins?

    d. Complete the equation: $2 + (-4) = $ _____.

2. Use coins to find the following sums. Make a sketch of your work. You may use more than six coins.

    a. $1 + (-5)$ b. $6 + (-4)$ c. $3 + (-3)$ d. $-5 + (-2)$

Use the coin model to answer the following questions.

3. a. Is the sum of two negative numbers positive or negative?

    b. How can you determine the sum of two negative numbers without using coins?

4. When is the sum of a positive and a negative number

    a. equal to 0?   b. positive?   c. negative?

5. How can you determine the sum of a positive and a negative number without using coins?

6. Use your rules from Exercises 3 and 5 to compute the following:

    a. $-17 + 25$ b. $13 + (-7)$ c. $-36 + (-19)$ d. $-11 + 11$

Activity 3: Subtraction Patterns

PURPOSE	Explore patterns to develop a rule for subtracting integers.
COMMON CORE SMP	SMP 1, SMP 2, SMP 7
GROUPING	Work individually.
GETTING STARTED	In each of the following sets of problems, complete the differences you know, then look for patterns to fill in the missing entries.

1. 4 – 0 = 4
    4 – 1 = 3
    4 – 2 = ____
    4 – 3 = ____
    4 – ____ = ____
    4 – ____ = ____
    4 – ____ = ____
    ____ – ____ = ____

2. 3 – 4 = −1
    2 – 4 = −2
    1 – 4 = ____
    0 – 4 = ____
    ____ – 4 = ____
    ____ – ____ = ____
    ____ – ____ = ____
    ____ – ____ = ____

3. 4 – 3 = 1
    4 – 2 = 2
    4 – 1 = ____
    4 – ____ = ____
    4 – ____ = ____
    4 – ____ = ____
    ____ – ____ = ____
    ____ – ____ = ____

4. −4 – 3 = −7
    −4 – 2 = ____
    −4 – 1 = ____
    −4 – ____ = ____
    ____ – ____ = ____
    ____ – ____ = ____
    ____ – ____ = ____

5. Next to each of the subtraction problems above, write a related addition problem using numbers that have the same absolute value as those in the given problem.

    Examples: $4 - 5 = 4 + (-5)$ $-4 - 1 = -4 + (-1)$

6. Write a rule for the subtraction of integers.

EXTENSION	Write problem situations that illustrate (1) subtraction of a negative integer from a positive integer and (2) subtraction of a negative integer from a negative integer. Write two problems for each case.

Activity 4: A Clown on a Tightrope

PURPOSE	Develop rules for addition and subtraction of integers using a number line model.
COMMON CORE SMP	SMP 2, SMP 4, SMP 5
MATERIALS	Other: A transparent copy of the large clown at the left
GROUPING	Work individually or in pairs.
GETTING STARTED	The clown performs his tightrope act according to the following rules:

- He starts each act standing on the first number.
- For addition, the clown faces right.
- For subtraction, he faces left.
- A positive second number tells the clown to walk forward.
- A negative second number tells him to walk backward.

Example:

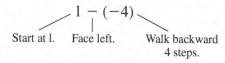

Start at 1.   Face left.   Walk backward 4 steps.

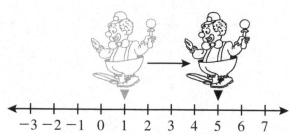

1.   Where does the clown's act end in the Example? What does $1 - (-4)$ equal?

Use the tightrope below and the transparent copy of the clown to solve each problem. The arrow should always point to the clown's location on the rope.

2. $2 + 3 =$ _____   3. $3 + (-5) =$ _____   4. $-4 + (-2) =$ _____

5. $-7 + 9 =$ _____   6. $3 - 7 =$ _____   7. $6 - (-2) =$ _____

8. $-5 - 3 =$ _____   9. $3 - (-6) =$ _____   10. $-2 - (-4) =$ _____

11. $-5 + 3 - (-8) =$ _____   12. $2 - (-7) + (-12) =$ _____

13. $-9 - (-11) - (-7) =$ _____

Use your results in Exercises 2–5 and the number line model to answer Exercises 14–16.

14. a. Is the sum of two negative integers positive or negative?

    b. How can you find the sum of two negative integers without using the number line model?

15. When is the sum of a positive and a negative integer

    a. positive?

    b. negative?

    c. equal to 0?

16. How can you determine the sum of a positive and a negative integer without using the number line model?

17. a. Write an addition problem in which the clown starts on the same number (3) and ends at the same answer as in Exercise 6.

    b. How are the numbers in the addition problem related to the numbers in the original problem?

18. Repeat Exercise 17 for each problem in Exercises 7–10.

19. Write a related addition problem that has the same answer as $-6 - (-4)$.

Activity 5: Multiplication and Division Patterns

PURPOSE   Explore patterns to develop rules for multiplication of integers and inverse operations to develop rules for division of integers.

COMMON CORE SMP   SMP 1, SMP 2, SMP 7

GROUPING   Work individually.

GETTING STARTED   In each of the following sets of problems, complete the products you know, then look for patterns to fill in the missing entries.

1. $4 \times 3 = 12$
    $4 \times 2 = \underline{\quad}$
    $4 \times 1 = \underline{\quad}$
    $4 \times \underline{\quad} = \underline{\quad}$
    $4 \times \underline{\quad} = \underline{\quad}$
    $\underline{\quad} \times \underline{\quad} = \underline{\quad}$
    $\underline{\quad} \times \underline{\quad} = \underline{\quad}$
    $\underline{\quad} \times \underline{\quad} = \underline{\quad}$

2. $4 \times 5 = 20$
    $3 \times 5 = 15$
    $\underline{\quad} \times 5 = 10$
    $\underline{\quad} \times 5 = \underline{\quad}$
    $\underline{\quad} \times 5 = \underline{\quad}$
    $\underline{\quad} \times \underline{\quad} = \underline{\quad}$
    $\underline{\quad} \times \underline{\quad} = \underline{\quad}$
    $\underline{\quad} \times \underline{\quad} = \underline{\quad}$

3. Write a rule for multiplying a positive number and a negative number.

4. $-3 \times 2 = -6$
    $-3 \times 1 = \underline{\quad}$
    $-3 \times 0 = \underline{\quad}$
    $-3 \times \underline{\quad} = \underline{\quad}$
    $-3 \times \underline{\quad} = \underline{\quad}$
    $\underline{\quad} \times \underline{\quad} = \underline{\quad}$
    $\underline{\quad} \times \underline{\quad} = \underline{\quad}$

5. $3 \times -6 = \underline{\quad}$
    $\underline{\quad} \times -6 = -12$
    $\underline{\quad} \times -6 = -6$
    $\underline{\quad} \times \underline{\quad} = \underline{\quad}$
    $\underline{\quad} \times \underline{\quad} = \underline{\quad}$
    $\underline{\quad} \times \underline{\quad} = \underline{\quad}$

6. Write a rule for multiplying two negative numbers.

Recall that multiplication and division are *inverse operations*.

Example: $12 \div 4 = 3$, since $3 \times 4 = 12$. In general, $A \div B = C$ means that $C \times B = A$.

$-12 \div 4 = ?$ Think: $4 \times ? = -12$
                          So $? = -3$.

Use the inverse relationship between multiplication and division to compute the quotients of several pairs of integers. Use the results to write a rule for division of integers.

Activity 6: Integer × and ÷ Contig

PURPOSE	Reinforce multiplication and division of integers.
COMMON CORE SMP	SMP 2, SMP 7
MATERIALS	Other: Integer × and ÷ Contig Game Board (page 86), three blank cubes, and chips for markers
GROUPING	Work in groups of two to four players.
GETTING STARTED	Follow the rules below to play *Integer × and ÷ Contig*.

RULES FOR INTEGER × AND ÷ CONTIG

1. Make three number cubes labeled as follows.　$-1, 2, 3, -4, -5, -6$

$$1, -2, 3, 4, -5, -6$$

$$1, -2, -3, -4, -5, 6$$

To begin play, place a chip on the **FREE** square on the game board.

2. Each player rolls the number cubes and finds the sum of the numbers showing on them. The player with the LEAST sum begins; play progresses to the next player on the right.

3. On each turn, the player rolls the number cubes and performs any combination of multiplication and/or division with the numbers showing on the cubes. The player then places a chip on the resulting number on the game board. A player may not place a chip on a number that is already covered.

4. To score points, a player must place a chip on a number on the board that is adjacent vertically, horizontally, or diagonally to a previously covered square. One point is scored for each adjacent covered square.

5. If a player is unable to produce a number that is not covered, he or she must pass the number cubes to the next player. If another player knows a play that can be made with the numbers on the number cubes, that player may call attention to the mistake and tell the other players the operations that will result in an uncovered number on the game board. The player citing the mistake may then place a chip on that number and earn points. This does not affect the turn of the player citing the mistake. If more than one player calls attention to a mistake, the first player to do so makes the play.

6. Players keep a running total of their scores. A player who cannot produce an uncovered number in three successive turns is eliminated from the game. When the game board is filled, or if all players have failed to play in three successive turns, the game ends. The player with the highest score is the winner.

Integer × and ÷ Contig

Game Board

–120	–90	–72	–48	–60	–25	–50
–32	–20	–36	–24	–30	–27	–40
–10	–12	–15	–6	–9	–8	–18
–1	3	–2	FREE	4	–3	1
18	8	6	30	12	9	10
50	54	48	60	72	90	120
64	40	36	25	20	24	15

Chapter Summary

In your early mathematical experiences, you probably thought about whole numbers as representing simple quantities, such as six marbles or five pencils. Thus the whole number 6 could be modeled by a set containing six objects. However, when your understanding of whole numbers was extended to the integers, your concept of number had to change in order to accommodate negative integers.

Like whole numbers, integers do represent quantities. However, when you think about an integer, you usually think of it as representing not only a quantity, but also a direction. Thus you think of 5 and –5 as opposites, as a $5 profit and a $5 loss, or as 5 more than zero and 5 less than zero. This interpretation distinguishes integers from whole numbers and is reflected in the models used to represent integers. On a number line, 5 and –5 are both located 5 units from zero, but –5 is to the left of zero and 5 is to the right. When modeled as particles, in their simplest forms 5 and –5 are represented by the same number of particles, but the particles have opposite charges. These ideas were explored in Activities 1 and 2.

The extension of whole numbers to integers required not only that you alter your concept of a number, but also that you modify your interpretations of the operations with numbers. Addition could still be thought of in terms of the union of sets. However, because the objects in the sets might be opposites, you found that in some cases you had to pair the opposites in order to find the sum. Similarly, by analyzing addition and subtraction using the number line model, you discovered that subtraction of integers can be interpreted as adding the opposite.

These changes in the meanings of addition and subtraction and the resulting algorithms were explored in Activities 2 and 4. The results were verified in Activity 3 by examining patterns. In Activity 5, patterns were analyzed to discover an algorithm for multiplying integers. The algorithm was extended to division by applying the inverse to find missing factors. The game in Activity 6 provided an opportunity to apply these algorithms.

Chapter 6
Fractions and
Rational Numbers

"Students should build their understanding of fractions as parts of a whole and as division. They will need to see and explore a variety of models of fractions, focusing primarily on familiar fractions such as halves, fourths, fifths, sixths, eighths, and tenths. ...They should develop strategies for ordering and comparing fractions, often using benchmarks such as $\frac{1}{2}$ or 1."

—Principles and Standards for School Mathematics

"Students apply their understanding of fractions and fraction models to represent the addition and subtraction of fractions with unlike denominators as equivalent calculations with like denominators. They develop fluency in calculating sums and differences of fractions, and make reasonable estimates for them. Students also use the meaning of fractions, of multiplication and division, and the relationship between multiplication and division to understand and explain why the procedures for multiplying and dividing fractions make sense."

—Common Core Standards for Mathematics

Other than whole-number computation, no topic in the elementary mathematics curriculum demands more time than the study of fractions. Yet, despite years of study, most students enter high school with a poor conception of fractions and an even poorer understanding of the operations with fractions. When asked about their memories of fractions, adults will often reply, *"Yours is not to reason why, just invert and multiply."*

Rational numbers should be taught as the natural extension of the whole numbers. The fraction $\frac{3}{4}$ can be viewed as the solution to the problem of dividing 3 dollars among 4 people, $3 \div 4$, or answering the question, "How many fourths (quarters) will each person receive?" For students to understand this connection between whole numbers and fractions, teaching about fractions and their operations should begin with concrete models, as was the instruction with whole numbers. A well-developed concept of fractions and a feeling for their magnitude must be established before they can be compared and ordered meaningfully. This must precede computation.

Number sense involving fractions and a deeper understanding of the operations with fractions must be developed prior to formal work with algorithms for the operations. This chapter provides activities which firmly develop the concepts of fractions. All operations are explored through a variety of concrete models and reinforced by representing the models pictorially.

Correlation of Chapter 6 Activities to the
Common Core Standards of Mathematical Practice

Activity Number and Title	Standards of Mathematical Practice
1: What Is a Fraction?	SMP 4, SMP 5
2: Square Fractions	SMP 1, SMP 4, SMP 5
3: Equivalent Fractions	SMP 4, SMP 5
4: How Big Is It?	SMP 2, SMP 3, SMP 6
5: What Comes First?	SMP 2, SMP 7
6: Fraction War	SMP 2, SMP 3
7: Adding and Subtracting Fractions	SMP 2, SMP 5, SMP 6
8: Multiplying Fractions	SMP 2, SMP 5, SMP 6
9: Dividing Fractions	SMP 2, SMP 5, SMP 6

Activity 1: What Is a Fraction?

PURPOSE	Develop the concept of a fraction.
COMMON CORE SMP	SMP 4, SMP 5
MATERIALS	Pouch: Pattern Blocks and Cuisenaire™ Rods
GROUPING	Work individually or in pairs.

Use pattern blocks to solve the following problems.

1. The trapezoid is what fractional part of the hexagon?   _____

2. The blue rhombus is what fractional part of the hexagon?   _____

3. The triangle is what fractional part of the hexagon?   _____

4. The triangle is what fractional part of the blue rhombus?   _____

5. The triangle is what fractional part of the trapezoid?   _____

What fractional part of each figure is *shaded*? *unshaded*?

1. 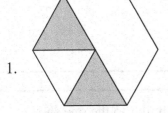   shaded _____   unshaded _____

2. 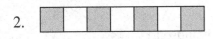   shaded _____   unshaded _____

3. 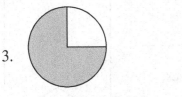   shaded _____   unshaded _____

4. 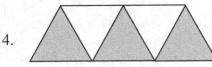   shaded _____   unshaded _____

Use Cuisenaire rods to solve the following.

1. If the orange rod = 1, each rod is what fractional part of the orange rod?

   a. red _____   b. green _____

   c. yellow _____   d. purple _____

2. If the purple rod = 1, each rod is what fractional part of the purple rod?

   a. brown _____   b. orange _____

   c. dark green _____   d. black _____

---

1. If the red rod = $\frac{1}{2}$, which rod = 1?   _____

2. If the red rod = $\frac{1}{3}$, which rod = 1?   _____

3. If the white rod = $\frac{1}{5}$, which rod = 1?   _____

4. If the white rod = $\frac{1}{4}$, which rod = $1\frac{3}{4}$?   _____

5. If the red rod = $\frac{1}{2}$, which rod = $1\frac{1}{2}$?   _____

6. If the red rod = $\frac{1}{3}$, which rod = $1\frac{2}{3}$?   _____

Explain how you used the rods to arrive at your answers.

Use two red trapezoids and one blue rhombus to construct a shape similar to the one shown below.

1. Given that the shape = 1, what pattern block(s) would you use to represent each of the following fractions?

   a. $\frac{1}{4}$ _____   b. $\frac{1}{2}$ _____   c. $\frac{1}{8}$ _____

Fill in the same shape using one red trapezoid, two blue rhombuses, and one green triangle.

2. What fraction is represented by each of the following?

   a. a blue rhombus _____   b. a red trapezoid _____   c. a green triangle _____

Activity 2: Square Fractions

PURPOSE	Reinforce the concept of a fraction and the meaning of equivalent fractions, and illustrate operations with fractions using geometric models.
COMMON CORE SMP	SMP 1, SMP 4, SMP 5
MATERIALS	Other: Sheet of paper 20 cm square (colored construction paper works well) and scissors (optional)
GROUPING	Work individually.
GETTING STARTED	Work through each section of the activity in order, following the folding and cutting directions carefully. If scissors are not used, fold and crease the paper sharply so that it will tear cleanly.

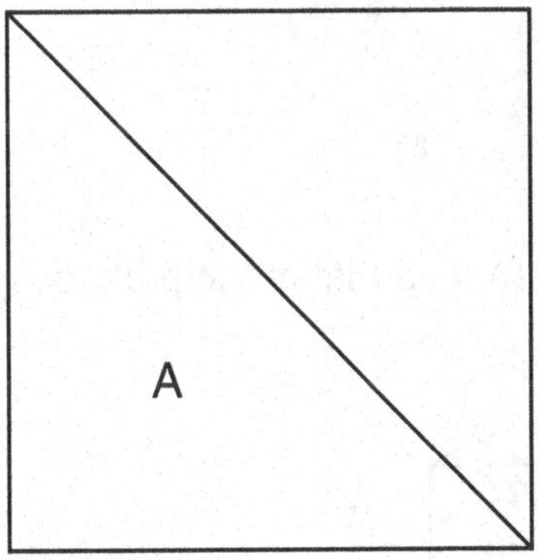

Fold the square as shown and cut or tear along the fold to divide the square into two congruent parts.

1. Each triangle is what fraction of the original square? _____

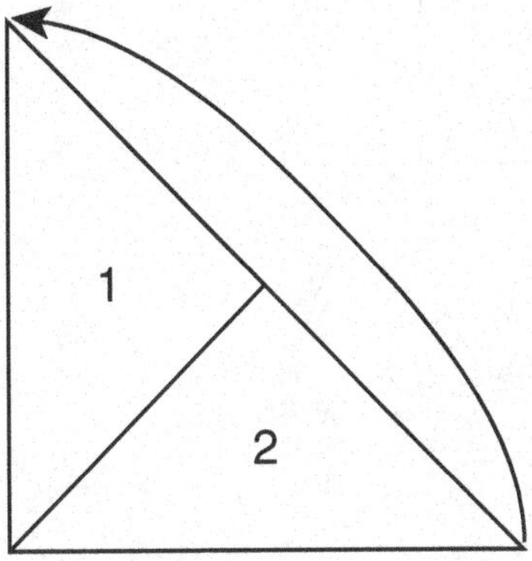

Pick one of the two triangles and fold it as shown. Cut or tear along the fold and label the two triangles 1 and 2.

2. Triangle 1 is what fractional part of

    a. triangle A? _____

    b. the original square? _____

    Explain your answers.

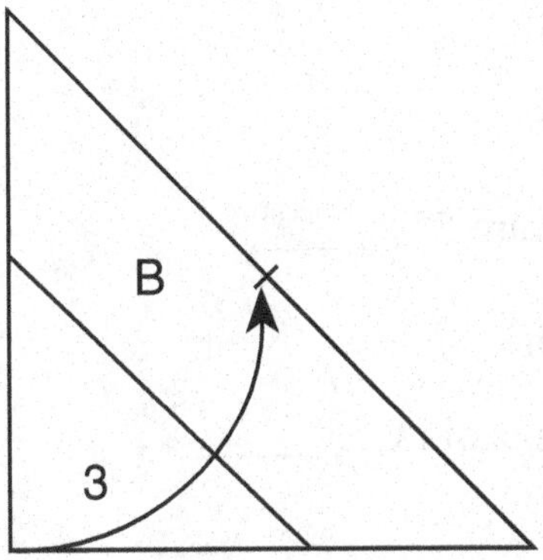

In the remaining large triangle, fold the vertex of the right angle to the midpoint of the longest side. Cut along the fold and label the polygons B and 3 as shown.

3. Triangle 3 is what fractional part of

    a. triangle 1? _____

    b. triangle A? _____

    c. trapezoid B? _____

Fold one of the endpoints of the longest side of trapezoid B to the midpoint of that side as shown. Cut along the fold and label the polygons C and 4.

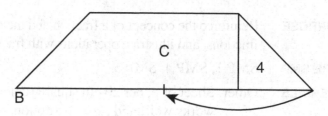

4. Triangle 4 is what fractional part of

    a. triangle 3? _____

    b. triangle A? _____

    c. trapezoid C? _____

Fold trapezoid C as shown. Cut along the fold and label the polygons D and 5.

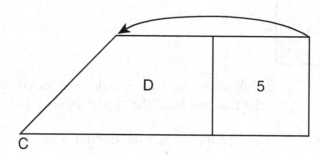

5. Square 5 is what fractional part of

    a. trapezoid C? _____

    b. trapezoid D? _____

    c. triangle 3? _____

    d. the original square? _____

6. Trapezoid C is what fractional part of

    a. triangle A? _____   b. square 5? _____

7. Trapezoid D is what fractional part of

    a. triangle 1? _____   b. trapezoid C? _____

Fold trapezoid D as shown. Cut along the fold and label the two polygons 6 and 7.

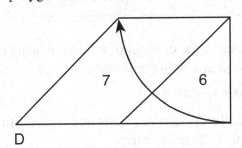

D

8. Triangle 6 is what fractional part of

   a. trapezoid D? _____

   b. triangle 3? _____

9. Parallelogram 7 is what fractional part of

   a. square 5? _____   b. trapezoid B? _____

   c. triangle 1? _____   d. original square? _____

10. Explain how triangle 1, triangle A, and the original square can be used to illustrate that $\dfrac{2}{4}$ is equivalent to $\dfrac{1}{2}$.

11. Explain how these figures can be used to illustrate that $\dfrac{1}{2}$ of $\dfrac{1}{2}$ is $\dfrac{1}{4}$, which is the multiplication problem $\dfrac{1}{2} \times \dfrac{1}{2} = \dfrac{1}{4}$.

12. If the original square is one unit, explain how trapezoid D and triangle 3 can be used to model the addition problem $\dfrac{3}{16} + \dfrac{1}{8} = \dfrac{5}{16}$.

13. If trapezoid B is one unit, explain how triangle 4 and trapezoid D can be used to model the division problem $\dfrac{1}{2} \div \dfrac{1}{6} = 3$.

Activity 3: Equivalent Fractions

PURPOSE   Develop the concept of a fraction using concrete models and a
problem-solving approach.

COMMON CORE SMP   SMP 4, SMP 5

MATERIALS   Pouch: Cuisenaire™ Rods and Pattern Blocks
Online: Fraction Strips

GROUPING   Work individually or in pairs.

For the following problems, use Cuisenaire rods to construct the trains.

1. Make all of the possible one-color trains the same length as a dark green rod and
complete the following. If dark green = 1, then

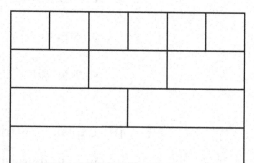

   a. light green = $\dfrac{1}{2}$ = $\dfrac{}{6}$

   b. red = $\dfrac{1}{}$ = $\dfrac{}{6}$

   c. purple = $\dfrac{}{}$ = $\dfrac{}{}$

   d. dark green = $\dfrac{}{6}$ = $\dfrac{}{}$

2. Make all of the possible one-color trains the same length as a brown rod and complete
each of the following. If brown = 1, then

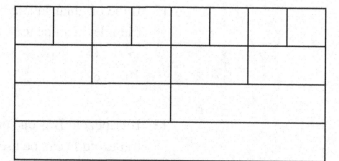

   a. purple = $\dfrac{}{}$ = $\dfrac{}{}$ = $\dfrac{}{}$

   b. dark green = $\dfrac{}{}$ = $\dfrac{}{}$

   c. red = $\dfrac{}{}$ = $\dfrac{}{}$

Use pattern blocks to construct a shape similar to the star and complete the following.

If the star shape $=1$, then

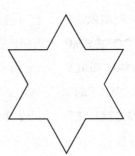

1. trapezoid $= \dfrac{1}{12} = \dfrac{1}{\quad}$

2. 2 blue rhombuses $= \dfrac{\quad}{6} = \dfrac{1}{\quad} = \dfrac{\quad}{\quad}$

3. hexagon $= \dfrac{6}{\quad} = \dfrac{\quad}{6} = \dfrac{1}{\quad}$

Find the fraction strips that can be folded into parts so that the resulting strip is equal in length to the fraction given in each problem. Folds may be made **only** on the lines on the fraction strips.

Write the name of the equivalent fractions in the space provided.

1. $\dfrac{1}{2} = \underline{\quad\quad} = \underline{\quad\quad} = \underline{\quad\quad} = \underline{\quad\quad}$

2. $\dfrac{2}{3} = \underline{\quad\quad} = \underline{\quad\quad} = \underline{\quad\quad}$

3. $\dfrac{3}{4} = \underline{\quad\quad} = \underline{\quad\quad}$

EXTENSIONS

1. Given a set of fractions equivalent to the fraction $\dfrac{a}{b}$, where $\dfrac{a}{b}$ is in lowest terms, what is the relationship among the set of numerators? the set of denominators?

2. Given a set of equivalent fractions, $\dfrac{a}{b}, \dfrac{c}{d}, \dfrac{e}{f}, \ldots$, where $\dfrac{a}{b}$ is in lowest terms, what is the relationship between the numerator and denominator of $\dfrac{a}{b}$ and the numerator and denominator of any equivalent fraction?

Activity 4: How Big Is It?

PURPOSE	Develop the ability to estimate the magnitude of a fraction.
COMMON CORE SMP	SMP 2, SMP 3, SMP 6
MATERIALS	Online: Fraction Sorting Board and Fraction Cards
GROUPING	Work in pairs.
GETTING STARTED	Use these rules to complete the following problems.

A fraction is close to

1   if the numerator and denominator are approximately the same size.

$\frac{1}{2}$ if the denominator is about twice as large as the numerator.

0   if the numerator is very small compared to the denominator.

THE FRACTION SORTING GAME

This is a game for two players. Cut out the fraction cards and shuffle them. Students take turns placing a card in the appropriate column on the fraction sorting board and justifying each placement to the other player. Reshuffle the deck and play again.

---

1. Complete the following fractions so that they are close to, but less than, $\frac{1}{2}$.

    a. $\dfrac{}{100}$ b. $\dfrac{}{25}$ c. $\dfrac{}{9}$ d. $\dfrac{}{14}$

    e. $\dfrac{7}{}$ f. $\dfrac{3}{}$ g. $\dfrac{11}{}$ h. $\dfrac{8}{}$

2. Complete the following fractions so that they are close to, but less than, 1.

    a. $\dfrac{}{27}$ b. $\dfrac{}{12}$ c. $\dfrac{}{75}$ d. $\dfrac{}{8}$

    e. $\dfrac{9}{}$ f. $\dfrac{3}{}$ g. $\dfrac{11}{}$ h. $\dfrac{95}{}$

EXTENSIONS

1. Given the fraction $\dfrac{13}{\square}$, what numbers would be acceptable in place of $\square$ so that the resulting fraction is *close to but less than 1*? Justify your answer.

2. Given the fraction $\dfrac{\square}{23}$, what numbers would be acceptable in place of $\square$ so that the resulting fraction is *close to $\frac{1}{2}$*? Justify your answer.

Activity 5: What Comes First?

PURPOSE	Compare and order fractions.
COMMON CORE SMP	SMP 2, SMP 7
MATERIALS	Pouch: Cuisenaire™ Rods and Pattern Blocks
	Online: Fraction Arrays
GROUPING	Work individually.

Use Cuisenaire rods to build all possible one-color trains that are the same length as a brown rod, and complete the following.

1. Which is larger, $\frac{5}{8}$ or $\frac{1}{2}$? _____   Complete the inequality: _____ > _____

2. Which is smaller, $\frac{3}{8}$ or $\frac{1}{4}$? _____   Complete the inequality: _____ < _____

3. Which is larger, $\frac{7}{8}$ or $\frac{3}{4}$? _____   Complete the inequality: _____ > _____

Use pattern blocks to construct a star shape (see Activity 3) and complete the following.

1. Which is larger, $\frac{5}{12}$ or $\frac{1}{2}$? _____   Complete the inequality: _____ < _____

2. Which is smaller, $\frac{2}{3}$ or $\frac{7}{12}$? _____   Complete the inequality: _____ > _____

3. Which is larger, $\frac{5}{6}$ or $\frac{3}{4}$? _____   Complete the inequality: _____ < _____

Use the Fraction Arrays to order the following sets of fractions.

1. $\frac{1}{2}, \frac{3}{5}, \frac{4}{7}$ _____ > _____ > _____

2. $\frac{2}{3}, \frac{3}{4}, \frac{7}{8}$ _____ > _____ > _____

3. $\frac{5}{12}, \frac{2}{5}, \frac{3}{7}$ _____ < _____ < _____

4. $\frac{5}{6}, \frac{11}{12}, \frac{4}{5}$ _____ < _____ < _____

Activity 6: Fraction War

PURPOSE	Reinforce estimation and comparison of fractions in a game format.
COMMON CORE SMP	SMP 2, SMP 3
MATERIALS	Online: Fraction Arrays and Fractions Game Board Other: Deck of playing cards (remove face cards)
GROUPING	Work in pairs.
GETTING STARTED	Follow the rules below to play *Fraction War*.

Example:

FRACTIONS GAME BOARD

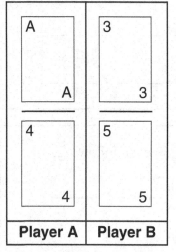

Player A wins a round in Game a: *the smaller fraction* is the winner.

Note:

When necessary, use the Fraction Arrays to compare the fractions.

1. Shuffle the cards and deal them evenly, face down to each player. Players choose a goal for a game from those listed below.

    a. Form a proper fraction by placing the card with the smaller number in the numerator. Player with the smaller fraction is the winner.

    b. Form a proper fraction by placing the card with the smaller number in the numerator. Player with the larger fraction is the winner.

    c. Form a proper fraction by placing the card with the smaller number in the numerator. Player with the fraction whose value is closest to $\frac{1}{2}$ is the winner.

    d. Form a fraction by placing the card with the larger number in the numerator. Player with the larger fraction is the winner.

    e. Place the first card in the numerator, the second in the denominator. Player with the fraction whose value is closest to 2 is the winner.

    f. Each player decides where to place each card. Player with the fraction whose value is closest to 1 is the winner.

2. Each player turns up two cards from his or her pile and follows the directions for the chosen game to form a fraction on the Fractions Game Board. The ace represents 1.

3. The winner of each round collects the four cards and places them face up at the bottom of his or her pile of cards. If the fractions formed are equivalent, each player turns over two additional cards and forms a new fraction. The winner of the round gets all eight cards.

4. When the players have played all the face-down cards, the player with the most face-up cards is the winner of the game. Reshuffle the cards and choose a different game.

Activity 7: Adding and Subtracting Fractions

PURPOSE	Develop algorithms for adding and subtracting fractions.
COMMON CORE SMP	SMP 2, SMP 5, SMP 6
MATERIALS	Pouch: Pattern Blocks Online: Fraction Strips
GROUPING	Work individually.

If the yellow hexagon = 1, then the red trapezoid = $\frac{1}{2}$, the blue rhombus = $\frac{1}{3}$, and the green triangle = $\frac{1}{6}$. Use pattern blocks to solve the following:

1.   1 red + 3 green   =   1 red + 1 red   =   3 green + 3 green   = _____

$$\frac{1}{2} + \frac{3}{6} \quad = \quad \frac{1}{2} + \frac{1}{2} \quad = \quad \frac{3}{6} + \frac{3}{6} \quad = \underline{}$$

2.   1 red + 1 blue   =   ___ green + ___ green   =   _____

$$\frac{1}{2} + \frac{1}{3} \quad = \quad \underline{} + \underline{} \quad = \quad \underline{}$$

3. 1 blue + 1 green   =   _____ + _____   =   _____

$$\frac{1}{3} + \frac{1}{6} \quad = \quad \underline{} + \underline{} \quad = \quad \underline{} \quad = \underline{}$$

4.   1 red − 1 blue   =   _____ − _____   =   _____

$$\frac{1}{2} - \frac{1}{3} \quad = \quad \underline{} - \underline{} \quad = \quad \underline{}$$

5.   1 red − 1 green   =   _____ − _____   =   _____

$$\frac{1}{2} - \frac{1}{6} \quad = \quad \underline{} - \underline{} \quad = \quad \underline{} \quad = \underline{}$$

Use pattern blocks to solve the following problems. Write your answers in simplest form, that is, the number represented by the least number of blocks of the same color.

Let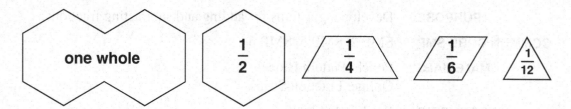

1. $\dfrac{1}{2} + \dfrac{3}{12} =$

2. $\dfrac{3}{4} + \dfrac{1}{2} + \dfrac{1}{6} =$

3. $\dfrac{3}{4} - \dfrac{2}{3} =$

4. $\dfrac{5}{6} - \dfrac{3}{4} =$

5. $\dfrac{2}{3} + \dfrac{1}{2} =$

6. $\dfrac{3}{4} + \dfrac{2}{3} + \dfrac{1}{6} =$

7. $1\dfrac{1}{2} + \dfrac{2}{3} =$

8. $1\dfrac{5}{12} - \dfrac{5}{6} =$

When adding fractions using fraction strips, you must fold the strips to show only the fractions that are needed. Strips are placed as shown in the following figures. A longer strip must then be found that has fold lines in common with the two fractions.

Example for Addition: $\dfrac{1}{3} + \dfrac{1}{4} =$

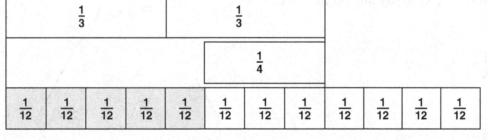

$\dfrac{1}{3}$ + $\dfrac{1}{4}$

$\dfrac{4}{12}$ + $\dfrac{3}{12}$ = $\dfrac{7}{12}$

Example for Subtraction: $\dfrac{2}{3} - \dfrac{1}{4} =$

$\dfrac{2}{3}$ − $\dfrac{1}{4}$ =

$\dfrac{8}{12}$ − $\dfrac{3}{12}$ = $\dfrac{5}{12}$

Use your fraction strips to solve the following problems.

1. $\dfrac{1}{2} + \dfrac{3}{5} =$ 2. $\dfrac{7}{8} - \dfrac{1}{4} =$ 3. $\dfrac{7}{10} - \dfrac{2}{5} =$

4. $\dfrac{5}{12} + \dfrac{1}{3} =$ 5. $\dfrac{2}{3} + \dfrac{3}{4} =$ 6. $\dfrac{3}{4} - \dfrac{1}{6} =$

EXTENSION   From your observations in this activity, write rules for adding and subtracting fractions with like and unlike denominators.

Activity 8: Multiplying Fractions

PURPOSE	Develop an algorithm for multiplying fractions.
COMMON CORE SMP	SMP 2, SMP 5, SMP 6
MATERIALS	Pouch: Pattern Blocks Other: Paper for folding
GROUPING	Work individually.

Example: $\frac{2}{3}$ of $\frac{1}{4}$ means two of three equal parts of $\frac{1}{4}$.

 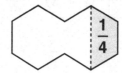

$$\frac{2}{3} \times \frac{1}{4} = \frac{2}{12} = \frac{1}{6}$$

Place pattern blocks on Figure A to solve the following multiplication problems. Record your solutions pictorially and numerically.

Figure A

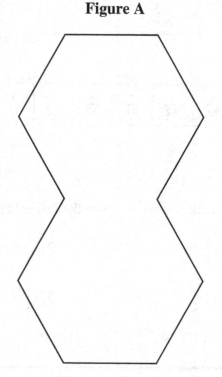

1. $\frac{1}{2} \times \frac{1}{3} =$

2. $\frac{3}{4} \times \frac{1}{3} =$

3. $\frac{1}{4} \times \frac{1}{3} =$

4. $\frac{3}{4} \times \frac{2}{3} =$

5. $\frac{5}{6} \times \frac{1}{2} =$

Example: $\frac{1}{2}$ of $\frac{2}{3}$ means one of the two equal parts of two thirds.

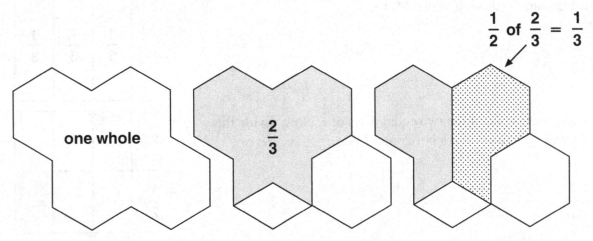

one whole

$\frac{2}{3}$

$\frac{1}{2}$ of $\frac{2}{3}$ = $\frac{1}{3}$

Use four hexagons to construct a figure similar to the one shown above, and solve the following problems. Record each step of your solutions both pictorially and numerically.

1. $\frac{3}{4} \times \frac{1}{6} =$

2. $\frac{3}{8} \times \frac{2}{3} =$

3. $\frac{7}{12} \times \frac{1}{2} =$

4. $\frac{5}{8} \times \frac{1}{3} =$

$\frac{1}{3}$ of 1 means one of the three equal parts of 1. Divide a piece of paper into thirds with vertical folds.

$\frac{1}{2}$ of $\frac{1}{3}$ means one of the two equal parts of $\frac{1}{3}$. Now, divide the thirds into halves with a horizontal fold.

$$\frac{1}{2} \text{ of } \frac{1}{3} = \frac{1}{6}$$

$\frac{2}{3}$ of $\frac{1}{2}$ means two of the three equal parts of $\frac{1}{2}$.

$$\frac{2}{3} \text{ of } \frac{1}{2} = \frac{2}{6} = \frac{1}{3}$$

Fold sheets of paper to solve the following multiplication problems. Record each step of your solutions pictorially and numerically.

1. $\dfrac{1}{3} \times \dfrac{3}{4} =$

2. $\dfrac{1}{4} \times \dfrac{2}{3} =$

3. $\dfrac{2}{3} \times \dfrac{2}{3} =$

4. $\dfrac{3}{8} \times \dfrac{1}{3} =$

EXTENSION   From what you have observed in this activity, write a rule for multiplication of fractions.

Activity 9: Dividing Fractions

PURPOSE	Develop understanding of division of fractions.
COMMON CORE SMP	SMP 2, SMP 5, SMP 6
MATERIALS	Pouch: Pattern Blocks
GROUPING	Work individually.
GETTING STARTED	Recall the use of the multiplication and division frame for the division of whole numbers.

Example: $3\overline{)6}$ can mean how many groups of 3 are there in 6?

In the following example ⬡ represents 1.

Example: $1 \div \frac{1}{2}$ means: How many groups of $\frac{1}{2}$ are there in 1?

$$1 \div \frac{1}{2} = 2$$

In each of the following, complete the sentence and then use pattern blocks to solve the related division problem.

1. $\frac{1}{3} \div \frac{1}{6}$ means _____

   $$\frac{1}{3} \div \frac{1}{6} =$$

2. $\frac{1}{2} \div \frac{1}{4}$ means _____

   $$\frac{1}{2} \div \frac{1}{4} =$$

3. $\frac{5}{6} \div \frac{5}{12}$ means _____

   $$\frac{5}{6} \div \frac{5}{12} =$$

4. $\frac{3}{4} \div \frac{1}{4}$ means _____

   $$\frac{3}{4} \div \frac{1}{4} =$$

5. $\frac{3}{2} \div \frac{3}{4}$ means _____

   $$\frac{3}{2} \div \frac{3}{4} =$$

To model the problem $\frac{1}{2} \div \frac{1}{3}$, let

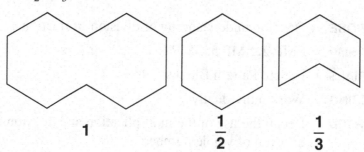

$$1 \qquad \frac{1}{2} \qquad \frac{1}{3}$$

How many sets of $\frac{1}{3}$ (two blue rhombuses) are there in $\frac{1}{2}$ (one hexagon)?

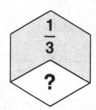

There is one group of $\frac{1}{3}$ (two blue rhombuses), plus a remainder.

The remainder is equal to one blue rhombus, which is $\frac{1}{2}$ of $\frac{1}{3}$.

Therefore $\frac{1}{2} \div \frac{1}{3} =$ one set of two blue rhombuses + one half set of two blue rhombuses

$$= \qquad 1 \qquad + \qquad \frac{1}{2}$$

$$= 1\frac{1}{2}$$

In each of the following, use pattern blocks to solve the division problems.

1. $\dfrac{5}{6} \div \dfrac{1}{3} =$

2. $\dfrac{3}{4} \div \dfrac{1}{2} =$

3. $\dfrac{2}{3} \div \dfrac{1}{2} =$

4. $1\dfrac{1}{3} \div \dfrac{1}{2} =$

EXTENSIONS   1. Describe how you would use fraction strips to solve the previous problems. Draw at least one illustration of your method.

2. From what you have observed in this activity, write a rule for division of fractions.

Chapter Summary

The activities in this chapter emphasized development of the conceptual understanding of rational numbers (fractions) and their operations. Concrete materials and structured lessons illustrated how the operations with fractions are an extension of the operations with whole numbers.

A curriculum developmentally appropriate for students is one of the central themes of the *Principles and Standards for School Mathematics* and the *Curriculum Focal Points for Prekindergarten through Grade 8 Mathematics*. Lessons should progress from the concrete to the representational (pictorial) to the abstract. The activities in this chapter illustrated such a curriculum through careful development of the operations with fractions.

Activity 1 developed the concept of a fraction using a variety of models. In different problems, you (a) determined the fractional part of a whole, (b) compared two areas or the length of two strips to determine a fraction, and (c) determined the whole unit given the fractional part.

In Activity 2, you used geometric representations of fractions that you constructed by folding and cutting a square. The pieces of the square helped you to explore the relationship among various pieces that made up the whole, and reinforced the understanding of addition, subtraction, multiplication, and division of fractions.

Activity 3 introduced equivalent fractions, a concept that is critical to ordering fractions, and adding and subtracting fractions. Activities 4–6 may be the most important ones in the chapter. They addressed another central theme of the *Principles and Standards for School Mathematics*: developing number sense. Only when the concept of a fraction is understood can one develop a sense of its size. Knowing terms like *about the same size*, *half as much*, and *very small as compared to* is critical when estimating the magnitude of a fraction.

Activity 7 used three models to explore addition and subtraction of fractions. Each one used the concept of equivalence as developed in previous activities. The importance of common denominators was connected to dividing the whole into equivalent parts that could then be added or subtracted. Activity 8 illustrated the language relationship between the word *of* and multiplication of rational numbers through a geometric model for fractions. Activity 9 modeled division of fractions in the same way that division of whole numbers was shown. That is, how many groups of one factor (the divisor) are there in the product (the dividend)? By modeling division this way, you can come to understand the familiar adage, "invert and multiply."

Chapter 7
Decimals, Real Numbers, and Proportional Reasoning

"Students understand decimal notation as an extension of the base-ten system of writing whole numbers that is useful for representing more numbers, including numbers between 0 and 1, between 1 and 2, and so on. Students relate their understanding of fractions to reading and writing decimals that are greater than or less than 1, identifying equivalent decimals, comparing and ordering decimals, and estimating decimal or fractional amounts in problem solving."

–Curriculum Focal Points for Prekindergarten through Grade 8 Mathematics

"Facility with proportionality involves much more than setting two ratios equal and solving for a missing term. It involves recognizing quantities that are related proportionally and using numbers, tables, graphs, and equations to think about the quantities and their relationship. Proportionality is an important integrative thread that connects many of the mathematics topics studied in grades 6–8."

—Principles and Standards for School Mathematics

"Percents, which can be thought about in ways that combine aspects of both fractions and decimals, offer students another useful form of rational numbers. Percents are particularly useful when comparing fractional parts of sets or numbers of unequal size, and they are also frequently encountered in problem-solving situations that arise in everyday life. As with fractions and decimals, conceptual difficulties need to be carefully addressed in instruction."

—Principles and Standards for School Mathematics

Instruction related to decimals and to computation with decimals will have its greatest impact when it is based on the same models and understanding as fractions and whole numbers. The activities in this chapter extend and reinforce the models that have been used previously with whole numbers and fractions. Decimal numbers will be modeled with base-ten blocks in a variety of ways.

Example: *If a rod = 1, then the small cube = 0.1.*

If a flat = 1, then a rod = 0.1, and the small cube = 0.01

111

These models reinforce the concept of a decimal number being part of a whole.

Number and operation sense and estimation skills with whole numbers developed in Chapters 2 and 3 will be reviewed and applied in the development of algorithms for multiplication of decimals.

Multiple representations will be used to introduce the concepts of ratio and proportion. In later chapters, ratio and proportion will be applied in real world contexts and used to explore proportional variation in geometry.

The concept of percent will be developed by extending rational number concepts and applying models that have been used previously with whole numbers, fractions, and decimals. The word percent means part of 100. The idea of part of a whole relates to the concept of a fraction. When the whole is 100, there is also a direct connection to decimals.

Correlation of Chapter 7 Activities to the Common Core Standards of Mathematical Practice

Activity Number and Title	Standards of Mathematical Practice
1: What's My Name?	SMP 2, SMP 5
2: Repeating Decimals	SMP 1, SMP 2, SMP 5, SMP 8
3: Race for the Flat	SMP 4, SMP 5
4: Empty the Board	SMP 1, SMP 4, SMP 5
5: Deci-Order	SMP 4, SMP 5, SMP 7
6: Decimal Arrays	SMP 3, SMP 4, SMP 5
7: Decimal Multiplication	SMP 3, SMP 4, SMP 5
8: Dice and Decimals	SMP 3, SMP 4, SMP 5
9: Professors Short and Tall	SMP 3, SMP 5, SMP 6
10: What Is Percent?	SMP 1, SMP 3, SMP 4, SMP 5

Activity 1: What's My Name?

PURPOSE	Develop the relationship between fractions and decimals.
COMMON CORE SMP	SMP 2, SMP 5
MATERIALS	Pouch: Base Ten Blocks
GROUPING	Work individually.
GETTING STARTED	Let 1 rod = 1, and 1 cube = 0.1.

Write the fraction and the decimal for the shaded part of each figure. If necessary, use the cubes to determine the decimal part.

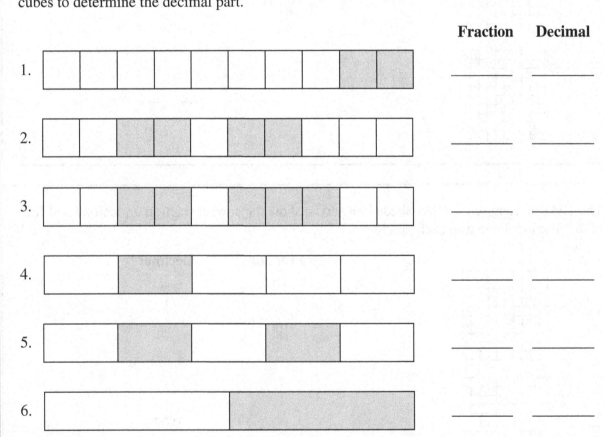

In the following problems, a flat = 1, a rod = 0.1, and a small cube = 0.01. Write the fraction and the decimal for the shaded part of each figure.

		Fraction	**Decimal**

1.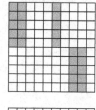

                                              _____   _____

2.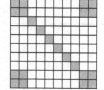

                                                _____   _____

3.

                                                _____   _____

The 100 grid represents a flat. Shade in a part to show the correct fraction or decimal and fill in the missing number in each problem.

		Fraction	**Decimal**

1.

                                        $\dfrac{63}{100}$ _____

2.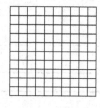

                                          _____   0.07

3.

                                          $\dfrac{14}{100}$ _____

Activity 2: Repeating Decimals

PURPOSE Apply the patterns problem-solving strategy to discover the decimal equivalent for proper fractions whose denominators contain only 9s or a combination of 9s followed by 0s.

COMMON CORE SMP SMP 1, SMP 2, SMP 5, SMP 8

MATERIALS Other: Calculator

GROUPING Work individually.

1. Convert each of the following fractions to its decimal equivalent. If your calculator rounds the last digit, ignore the rounded digit.

   a. $\dfrac{1}{9}$ = _____

   b. $\dfrac{7}{9}$ = _____

   c. $\dfrac{5}{9}$ = _____

   d. $\dfrac{3}{9}$ = _____

2. a. How many digits are there in the repeating block of the decimal equivalent of each fraction? _____

   b. How many 9s are there in the denominator of each fraction? _____

   c. What is the relationship between the numerator of each fraction and its decimal equivalent?

3. Convert each of the following fractions to its decimal equivalent.

   a. $\dfrac{13}{99}$ = _____

   b. $\dfrac{47}{99}$ = _____

   c. $\dfrac{5}{99}$ = _____

   d. $\dfrac{214}{999}$ = _____

   e. $\dfrac{75}{999}$ = _____

   f. $\dfrac{235}{9999}$ = _____

   g. $\dfrac{8457}{9999}$ = _____

   h. $\dfrac{53071}{99999}$ = _____

4. What is the relationship between the number of digits in the repeating block and the number of 9s in the denominator?

5. What is the relationship between the numerator of each fraction and its decimal equivalent?

6. Predict the decimal equivalent for each of the following fractions. Then use your calculator to check your predictions.

    a. $\dfrac{615}{999} =$ _____

    b. $\dfrac{8}{99} =$ _____

    c. $\dfrac{137}{9999} =$ _____

    d. $\dfrac{7421}{99999} =$ _____

7. Write a rule for finding the decimal equivalent of any proper fraction whose denominator contains only 9s.

8. Explain how the rule that you wrote above can be extended to include the decimal equivalent for an improper fraction such as $\dfrac{25}{9}$ or $\dfrac{137}{99}$.

9. a. Convert $\dfrac{1}{3}$ and $\dfrac{2}{3}$ to decimals. $\dfrac{1}{3} =$ _____ $\dfrac{2}{3} =$ _____.

    b. What is the sum of the decimal numbers? _____

    c. $\dfrac{1}{3} + \dfrac{2}{3} =$ _____

    d. What can you conclude from your answers to Parts b and c?

10. Use the fact that $\dfrac{25}{90} = \dfrac{25}{9} \times \dfrac{1}{10} = \left(2 + \dfrac{7}{9}\right) \times \dfrac{1}{10}$, to determine the decimal equivalent of $\dfrac{25}{90}$.

    $\dfrac{25}{90} =$ _____

11. Use the procedure from Exercise 10 to predict the decimal equivalent for each of the following fractions. Then use your calculator to check your predictions.

    a. $\dfrac{32}{90} =$ _____

    b. $\dfrac{76}{90} =$ _____

    c. $\dfrac{583}{900} =$ _____

    d. $\dfrac{347}{900} =$ _____

    e. $\dfrac{28}{990} =$ _____

    f. $\dfrac{2743}{9900} =$ _____

EXTENSION   1.   Complete the following table.

Fraction	Decimal Equivalent	Repeating or Terminating	Prime Factorization of Denominator	Number of Digits in the Repeating Block	Number of Non-Repeating Digits after the Decimal
$\dfrac{2}{3}$	$0.\overline{6}$	Repeats	3^1	1	0
$\dfrac{29}{90}$	$0.3\overline{2}$	Repeats	$2^1 \cdot 3^2 \cdot 5^1$	1	1
$\dfrac{17}{99}$	$0.\overline{17}$				
$\dfrac{3}{125}$		Terminates			3
$\dfrac{1}{8}$					
$\dfrac{17}{400}$		Terminates			
$\dfrac{119}{5000}$					
$\dfrac{4179}{4950}$					
$\dfrac{5}{24}$				1	
$\dfrac{121}{600}$					

2.   a.   What are the prime numbers in the prime factorization of the denominators of those fractions whose decimal equivalents terminate? _____

   b.   Does the prime factorization of the denominator of any fraction whose decimal equivalent repeats contain only the prime factors in Part a? If so, which fractions?

3.   When a decimal is converted to a fraction, the denominator will be 10 or _____ or _____ or _____ or _____, etc.

4.   Write the prime factorization of each power of 10.

   a. 10   b. 100   c. 1000   d. 10,000

5.   Based on your answers to Exercises 2 through 4, how can you predict whether the decimal equivalent of a fraction in lowest terms will terminate or repeat?

6.   For fractions in lowest terms whose decimal equivalents terminate, how is the number of digits in the decimal equivalent related to the exponents in the prime factorization of the denominator?

7.   For fractions in lowest terms whose decimal equivalents repeat, how can you predict the number of digits between the decimal point and the repeating block in the decimal equivalent?

Activity 3: Race for the Flat

PURPOSE   Model addition of decimals using Base Ten Blocks.

COMMON CORE SMP   SMP 4, SMP 5

MATERIALS   Pouch: Base Ten Blocks
Other: Number cubes labeled as shown

	.7		
.1	.3	.5	.2
	.8		

	.09		
.04	.02	.03	.05
	.07		

GROUPING   Work in pairs.

GETTING STARTED   In this game, the flat = 1, the rod = 0.1, and the cube = 0.01. Players alternate turns rolling the dice. After each roll, a player collects the correct number of rods and cubes corresponding to the numbers on the dice and places them on his or her flat. On each roll after the first, a player adds the new blocks to those on the flat, trading when necessary. Each addition must be recorded and validated by the blocks on the flat as shown below. The winner is the first player to cover the flat exactly. On any turn, a player may choose to roll only one die.

Example:

Turn 1   Turn 2

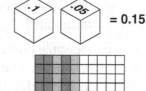 = 0.27   = 0.15

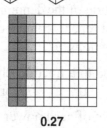   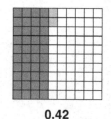

0.27   0.27 + 0.15   =   0.42

Play another game in which a player does not have to fill the flat exactly. Choose a goal for the game from one of the following:

A.   The winner is the person whose number is closest to 1.
B.   The winner is the person whose number is the greatest.

Players roll the dice and add blocks to the flat as above. For Goal A, a player's strategy will determine when to stop rolling the dice. For Goal B, players stop adding blocks on the roll that covers the flat, with or without some blocks left over.

At the conclusion of a game, players compare the decimal numbers represented by all of their blocks. The winner is the person whose number meets the goal of the game.

Activity 4: Empty the Board

PURPOSE	Model subtraction of decimals using Base Ten Blocks.
COMMON CORE SMP	SMP 1, SMP 4, SMP 5
MATERIALS	Pouch: Base Ten Blocks Online: Place-Value Operations Board Other: Tenths and hundredths decimal dice (see Activity 3) and a colored chip
GROUPING	Work in pairs.
GETTING STARTED	In this activity, a flat = 1. Place the chip on the place-value operations board as shown to represent the decimal point. Each player arranges a stack of five flats on his or her board. Players alternate turns rolling the dice. After each roll, the player who rolled the dice removes blocks equivalent to the sum of the numbers on the dice, making trades when necessary. Then the player must record the subtraction problem that was modeled by removing the blocks. The first player to empty the entire board is the winner.

Example:

PLACE-VALUE OPERATIONS BOARD

Roll 1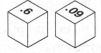

$$\begin{array}{r} 5.00 \\ -\ .69 \\ \hline 4.31 \end{array}$$

Roll 2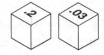

$$\begin{array}{r} 4.31 \\ -\ .23 \\ \hline 4.08 \end{array}$$

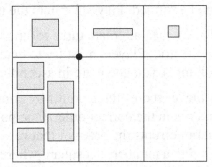

EXTENSION	How would you adapt this game to reinforce addition of decimals, and how would you adapt the game in Activity 3 for subtraction?

DECIMAL PUZZLE

In each of the given numbers, the decimal point may be placed in front of the first digit, behind the last digit, or between any two digits. Place the decimal point in each number so that the sum of the resulting numbers is 100.

$$\begin{array}{r} 6\ 8 \\ 9\ 5\ 9 \\ 4\ 8\ 1 \\ 1\ 7\ 3\ 4 \\ 2\ 2\ 7\ 9 \\ +\ 4\ 0\ 1\ 1 \end{array}$$

Answer is 100

Activity 5: Deci-Order

PURPOSE	Reinforce ordering of decimals.
COMMON CORE SMP	SMP 4, SMP 5, SMP 7
MATERIALS	Online: Deci-Order Grids
	Other: Two polyhedral dice (8 or 10 faces) or cubes marked as shown

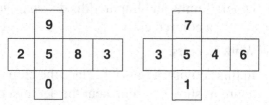

GROUPING	Work in pairs.
GETTING STARTED	Follow the rules below to play *Deci-Order*.

RULES FOR DECI-ORDER

1. Player 1 rolls the dice and forms a decimal number with the numbers showing on the dice. All numbers will be hundredths. For example, if 3 and 5 are rolled, 0.35 or 0.53 can be formed. Player 1 enters the number formed in any square on the grid.

2. Players then alternate turns rolling the dice, forming a decimal number, and placing it on the grid. Players must try to enter their numbers so that the entries in each row, column, or diagonal are in ascending or descending order.

3. A player scores three points by completing a row, column, or diagonal in which each entry is in the correct order. One point is lost if a player enters a number in a square that interrupts the order of that row, column, or diagonal. No points are lost for playing a number in a square where the order of the row, column, or diagonal is already interrupted.

   Examples:

   Deci-Order

.39	.56	.21
.95		.39
.51		
.62	.71	.85

   A. If a player rolls 7 and 5, two points would be scored by playing 0.75 in row 4, column 1 of the grid at the left, since three points would be scored for completing the diagonal and one point would be lost for interrupting the order of the row.

   B. If 0.57 were placed in that position instead of 0.75, six points would be scored for completing both the row and the diagonal.

   C. No points would be scored by placing 0.57 or 0.75 in row 1, column 1, since the order of row 1 was previously interrupted.

4. Players keep a running score after each turn; play continues until the grid is filled.

Activity 6: Decimal Arrays

PURPOSE	Develop an algorithm for multiplication of decimals.
COMMON CORE SMP	SMP 3, SMP 4, SMP 5
MATERIALS	Pouch: Base Ten Blocks
GROUPING	Work individually.
GETTING STARTED	Recall from Activity 8 in Chapter 6 on multiplication of fractions that 0.2×0.3 means 0.2 of 0.3. In this activity, a flat = 1, a rod = 0.1, and the small cube = 0.01.
Example:	To determine 0.2×0.3, place a flat on your paper and label the factors as shown.

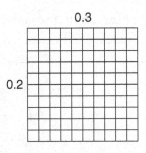

Place three rods vertically on the flat as shown to model 0.3.

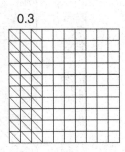

Place two rods horizontally as shown to model 0.2.

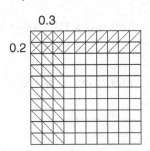

The product of 0.2×0.3 is determined by the rectangular array formed by the overlapping parts of the rods. The six squares in the array represent *6 parts of 100*.

So, 0.2 of 0.3 or $0.2 \times 0.3 = 0.06$.

Use base ten blocks to solve these multiplication problems. Record your results on the 100 grids by labeling the dimensions (factors) and shading in the correct number of rods. Find the number of squares in the overlapping area to determine each product.

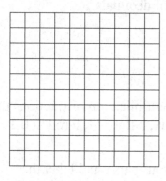   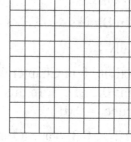   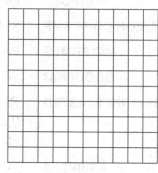

0.4 × 0.7 = _____   0.5 × 0.2 = _____   0.8 × 0.9 = _____

From what you have observed in this activity, write a rule for multiplying decimals.

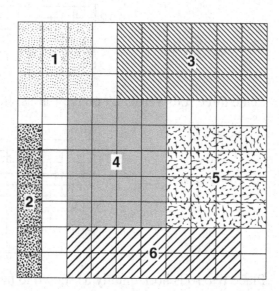

The 100 grid above represents a flat that is equal to 1. Write the dimensions for each shaded rectangle and determine its area.

Rectangle	Dimensions	Area		Rectangle	Dimensions	Area
1	___ × ___	___		2	___ × ___	___
3	___ × ___	___		4	___ × ___	___
5	___ × ___	___		6	___ × ___	___

How can you use the dimensions of the rectangle to determine its area?

Activity 7: Decimal Multiplication

PURPOSE	Use Base Ten Blocks to extend multiplication of decimals.
COMMON CORE SMP	SMP 3, SMP 4, SMP 5
MATERIALS	Pouch: Base Ten Blocks Online: Multiplication and Division Frame
GROUPING	Work individually.

For these problems, a flat = 1. Model the multiplication by using base ten blocks to construct a rectangle in the multiplication and division frame. Then determine the product of the two factors. Record your work as shown in the example.

Example: 2.4 × 1.3 =

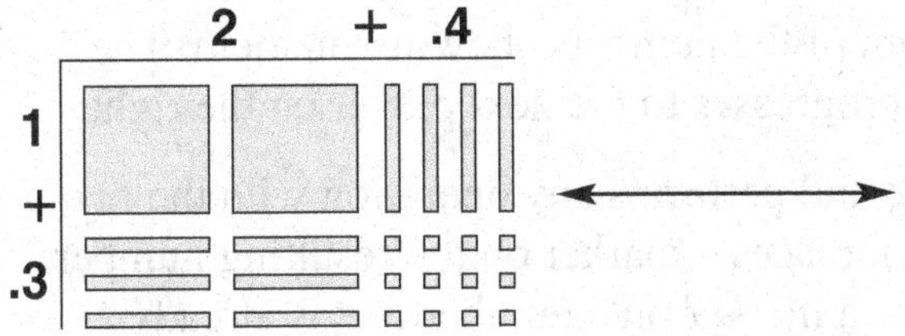

$$
\begin{array}{r}
2 + .4 \\
\times\ 1 + .3 \\
\hline
2 + .4 \\
.6 + .12 \\
\hline
2 + 1.0 + .12 = 3.12
\end{array}
$$

1. 3 × 2.3 = _____

2. 1.4 × 4.2 = _____

3. 3.4 × 1.2 = _____

4. 2.2 × 3.5 = _____

EXTENSION	Construct some rectangular arrays to represent the product of two decimals. Give the arrays to another student, identifying the correct decimal value for each base ten block. Instruct the student to determine the factors of your number.

Activity 8: Dice and Decimals

PURPOSE	Reinforce operations with decimal numbers.
COMMON CORE SMP	SMP 3, SMP 4, SMP 5
MATERIALS	Other: Dice and Decimals game board (page 125), a pair of dice, and chips for markers
GROUPING	Work in groups of two to four players.
GETTING STARTED	Follow the rules below to play *Dice and Decimals*.

RULES FOR DICE AND DECIMALS

1. The dots on each face of the dice represent a decimal number in tenths: 0.1, 0.2, 0.3, 0.4, 0.5, and 0.6. To begin play, place a chip on each **FREE** square on the game board.

2. Each player rolls the dice and finds the sum of the numbers showing on them. The player with the LEAST sum begins; play progresses to the next player on the right.

3. On his or her turn, the player rolls the dice and performs any operation with the two numbers showing on them. The player then places a marker on the resulting number on the game board. A player may not place a marker on a number that is already covered. When using division, a remainder is not allowed.

    Example: $0.6 \div 0.5 = 1.2$ can be played, but $0.5 \div 0.6 = 0.8\overline{3}$ is not allowed.

4. To score points, a player must place a marker on a number on the game board that is adjacent vertically, horizontally, or diagonally to a previously covered square. One point is scored for each adjacent covered square. No points are scored by placing a marker on a square that touches the **NO COUNT** square.

5. If a player is unable to produce an uncovered number, he or she must pass the dice to the next player. If another player knows a play that can be made with the numbers on the dice, that player may call attention to the mistake and tell the other players the operation that will result in an uncovered number. The player citing the mistake may then place a marker on that number and earn points. This does not affect the turn of the player citing the mistake. If more than one player calls attention to a mistake, the first player to do so makes the play.

6. Players keep a running total of their scores. A player who cannot produce an uncovered number in three successive turns is eliminated from the game. When the game board is filled, or if all players have failed to play in three successive turns, the game ends. The player with the highest score is the winner.

Dice and Decimals
Game Board

FREE	0	.01	.02	.03	.04	FREE
.05	.06	.08	.09	.1	.12	.15
.16	.18	.2	.24	.25	.3	.36
.4	.5	.6	NO COUNT	.7	.75	.8
.9	1	1.1	1.2	1.25	1.5	2
FREE	2.5	3	4	5	6	FREE

Activity 9: Professors Short and Tall

PURPOSE	Introduce the concepts of ratio and equivalent ratios.
COMMON CORE SMP	SMP 3, SMP 5, SMP 6
MATERIALS	Pouch: Colored Squares Other: 15 pennies
GROUPING	Work individually or with a partner.

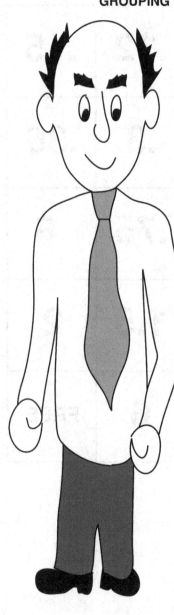

1. The man at the left is Professor Short.

    a. How many pennies tall is he? _____

    b. How many squares tall is he? _____

2. a. Professor Tall looks exactly like Professor Short, but he is 10 pennies tall! Without measuring, predict Professor Tall's height in squares. _____

    b. Explain your thinking.

3. a. Even though you do not have a picture of Professor Tall, you can still measure his height in squares. Explain how you could do this.

    b. How many squares tall is he? _____

    c. Is this the same as your prediction? If not, can you explain why?

A *ratio* is a comparison of two numbers or measures by division. The ratio of Professor Short's height to Professor Tall's height can be written in four ways.

$$8 \text{ to } 10 \qquad \frac{8}{10} \qquad 8 \div 10 \qquad 8 : 10$$

4. Write the ratio of Professor Short's height in squares to Professor Tall's height in squares in four different ways.

5. The following table contains the heights for some other people who look exactly like Professor Short.

    a. Fill in the missing numbers in the table.

Height in Pennies	4	8	12	16	20		40
Height in Squares		6				21	

    b. Explain how you found the missing numbers.

6. Use the information in the table in Exercise 5(a) to answer the following questions:

   a. If a person is 27 squares tall, what is his/her height in pennies? _____

   b. If a person is 10 pennies tall, what is his/her height in squares? _____

   c. How does each pair of numbers in the table appear to be related?

In the table in Exercise 5(a), the ratios of the heights in pennies to the heights in squares are equivalent. Ratios that can be expressed as equivalent fractions or quotients are *equivalent*.

7. Is the ratio of Professor Short's height in squares to Professor Tall's height in squares (the ratio you found in Exercise 4) equivalent to the ratio of their heights in pennies? Explain why or why not.

8. Graph the pairs of numbers in the table on the following coordinate grid.

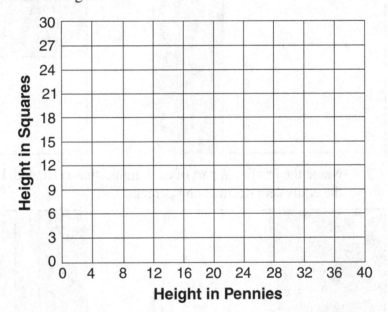

   a. What do you notice about the points?

   b. Use the graph to answer the following questions:

      If a person is 30 pennies tall, what is his/her height in squares? _____

      If a person is 18 squares tall, what is his/her height in pennies? _____

Activity 10: What Is Percent?

PURPOSE	Develop the concept of percent as a part of 100 and the relationship among fractions, decimals, and percents.
COMMON CORE SMP	SMP 1, SMP 3, SMP 4, SMP 5
MATERIALS	Online: A transparent copy of a 100 Grid
GROUPING	Work individually.
GETTING STARTED	Percent means *part of* 100. 40% means 40 equal parts out of 100. 80 correct out of 100 problems $= \frac{80}{100} = 80\%$.

1. Place a copy of the 100 grid over each square. Count the shaded squares in the grid to determine what part of 100 and the percent of each figure that is shaded.

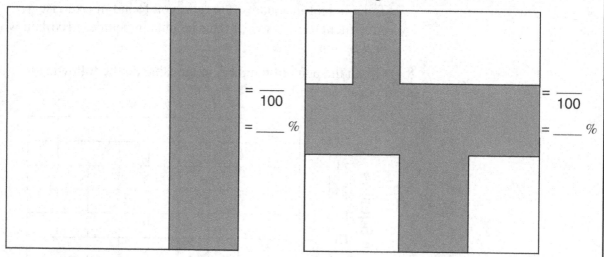

$= \dfrac{}{100}$

$= \underline{}\ \%$

$= \dfrac{}{100}$

$= \underline{}\ \%$

2. Name the fractional part of each figure that is shaded. Use your 100 grid to determine the equivalent decimal and percent.

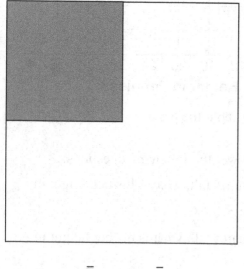

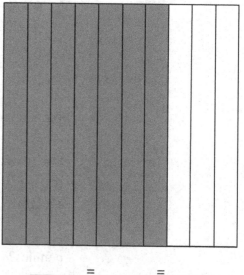

$\dfrac{}{\text{Fraction}} = \dfrac{}{\text{Decimal}} = \dfrac{}{\text{Percent}}$

$\dfrac{}{\text{Fraction}} = \dfrac{}{\text{Decimal}} = \dfrac{}{\text{Percent}}$

Complete the following.

1.

40% means ____ out of 100,
or 20 out of 50,
or ____ out of 25,
or 4 out of ____,
or 6 out of ____,
or 60 out of ____.

2.

15% means ____ out of 100,
or ____ out of 20,
or ____ out of $5.00,
or 12 out of ____,
or 24 out of ____,
or ____ out of $25.00.

1. Shade 20% of the squares.
$$\frac{20}{100} = \frac{}{15}$$

2. Shade 75% of the triangles.
$$\frac{75}{100} = \frac{}{8}$$

3. Shade 25% of the circles.
Circle 50% of the circles.

4. Put an X in 35% of the rectangles.
Put a Y in 25% of the rectangles.
Put a Z in 15% of the rectangles.

EXTENSION 1. List five real-world applications of percent between 28% and 57%.

2. List three real-world applications of 200%.

3. A headline on the newspaper business page reads, *ABC stock price increases 200% in the past six months from $7\frac{1}{2}$ to 15.* Do you believe the headline? Explain your answer.

Chapter Summary

The activities in this chapter stressed the conceptual development of decimals. The activities continued the constructivist approach to fractions in Chapter 6 and progressed from the concrete to the representational (pictorial) to the abstract or symbolic level.

Activity 1 used models to demonstrate the relationship between fractions and decimals. In Activity 2, you explored patterns in the decimal equivalents of fractions whose denominators contain only 9s or a combination of 9s followed by 0s. In addition, you learned how to determine whether the decimal equivalent of a fraction terminates or repeats.

The games in Activities 3 and 4 provided opportunities to physically place rods and cubes on a flat during the addition and subtraction process and then to make the necessary trades in order to represent the answer with the least number of blocks. This physical modeling of the operations and the trading and regrouping process are all essential for understanding the computational algorithms. The game in Activity 5 provided an opportunity to practice ordering of decimal numbers. The activity also involved problem solving in order to determine the maximum score in each play.

Activities 6 and 7 used the rectangular array model for multiplication to extend the processes developed with whole numbers and fractions to decimals. The array model will be revisited in the geometry section. Using this model in a variety of settings provides a firm foundation for understanding the concept of area.

A game format was also used in Activity 8 to provide practice in all operations with decimal numbers. In order to maximize the points earned on a play, problem-solving strategies were employed to determine all of the possible outcomes using the two numbers on the dice and then deciding which result produced the greatest score.

Activity 9 used multiple representations to introduce the concepts of ratio and equivalent ratios. The use of the 100 Grid in Activity 10 established the relationship between percent and 100 and also helped to develop the connections among fractions, decimals, ratios, and percents. These connections were applied in determining the percent of a number by finding a fractional part or multiplying by a decimal.

Chapter 8
Algebraic Reasoning, Graphing, and Connections with Geometry

"To think algebraically, one must be able to understand patterns, relations, and functions; represent and analyze mathematical situations and structures using algebraic symbols; use mathematical models to represent and understand quantitative relationships; and analyze change in various contexts."

—Navigating through Algebra in Grades 3–5

"... students should be able to understand the relationships among tables, graphs, and symbols and to judge the advantages and disadvantages of each way of representing relationships for particular purposes. As they work with multiple representations of functions— including numeric, graphic, and symbolic—they will develop a more comprehensive understanding of functions ..."

—Principles and Standards for School Mathematics

In this chapter, you will review basic concepts of algebra involving variables, expressions, and equations and extend your knowledge of these concepts to provide a foundation for algebraic thinking. Connections to geometry will be made by finding algebraic expressions that describe geometric patterns.

The concept of function is one of the central unifying themes in mathematics. The study of arithmetic, algebra, geometry, probability, and statistics relies on generalizing patterns and developing mathematical models (functions) that can be used to describe real-world situations. Through these activities, you will see how functions can be described as formulas, tables, and graphs, and examine multiple representations of the same function.

The use of a Cartesian coordinate system, named for its inventor René Descartes, makes it possible for mathematicians to link geometric and algebraic representations of problems. Coordinate geometry also has wide application in the real world. Finding a city on a state map, using longitude and latitude to determine a location on a global map, and graphing a system of algebraic equations to find the maximum earnings based on several variables that affect production of an item are just a few of the everyday uses of coordinate geometry.

131

**Correlation of Chapter 8 Activities to the
Common Core Standards of Mathematical Practice**

Activity Number and Title		Standards of Mathematical Practice
1:	Patterns and Expressions	SMP 1, SMP 4, SMP 5, SMP 7, SMP 8
2:	Regular Polygons in a Row	SMP 3, SMP 7, SMP 8
3:	What's My Function?	SMP 1, SMP 2, SMP 4, SMP 5
4:	Exploring Linear Functions	SMP 1, SMP 2, SMP 3, SMP 4, SMP 5, SMP 8
5:	Graphing Rectangles	SMP 1, SMP 2, SMP 3, SMP 4, SMP 5, SMP 6, SMP 7, SMP 8

Activity 1: Patterns and Expressions

PURPOSE	Write algebraic expressions that describe patterns.
COMMON CORE SMP	SMP 1, SMP 4, SMP 5, SMP 7, SMP 8
MATERIALS	Other: Colored pencils or markers
GROUPING	Work in pairs.
GETTING STARTED	Small colored tiles are used to tile larger squares with yellow tiles on the corners, red tiles along the edges, and blue tiles forming a square in the interior. The first two tiled squares are shown below.

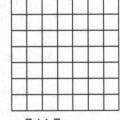

☐ = yellow

▨ = red

▇ = blue

3 × 3 square   4 × 4 square

1. Color the next three squares and record the number of tiles of each color in the table. The first two columns of the table have been done for you. Look for patterns that allow you to fill in the table for an 8 × 8 square.

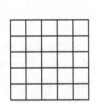 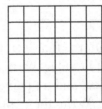 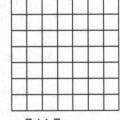

5 × 5 square   6 × 6 square   7 × 7 square

Dimensions of square	3 × 3	4 × 4	5 × 5	6 × 6	7 × 7	8 × 8	...	*n* × *n*
Number of yellow tiles	4	4					...	
Number of red tiles	4	8					...	
Number of blue tiles	1	4					...	

2. Look for patterns that lead to algebraic expressions for the number of yellow, red, and blue tiles in an $n \times n$ square. Write the expressions in the table.

3. What are the dimensions of the square made up of 400 colored tiles? How many tiles are there of each color?

4. Explain what each of the expressions in Exercise 2 represents physically with the tiles.

    Example: The blue tiles form a square; the length of its side is two less than n, the length of a side of the outside square. So the number of blue tiles $= (n - 2)^2$.

EXTENSION   1.   Unit cubes are used to build a larger cube that has edges 3 units long. Then the six faces of the larger cube are painted. When this cube is taken apart, how many of the unit cubes will have exactly 3 painted faces? 2 painted faces? 1 painted face? no painted faces?

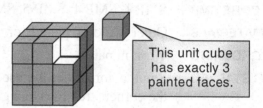

This unit cube has exactly 3 painted faces.

2.   Repeat Exercise 1 for cubes with edges 4 units and 5 units long. Record your results in the table.

Length of edge		3 units	4 units	5 units	...	*n* units
Number of unit cubes with	3 painted faces				...	
	2 painted faces				...	
	1 painted face				...	
	0 painted faces				...	

3.   Look for patterns that lead to expressions that represent the number of unit cubes with 3 painted faces, 2 painted faces, 1 painted face, and no painted faces in a cube with an edge *n* units long. Write the expressions in the table.

4.   a.   What is the length of an edge of a large cube made up of 729 unit cubes?

    b.   How many of the unit cubes will have 3 painted faces? 2 painted faces? 1 painted face? no painted faces?

5.   Explain what each of your expressions in Exercise 3 represents physically with the cubes.

Activity 2:   Regular Polygons in a Row

PURPOSE Reinforce the concept of perimeter and use the patterns problem-solving strategy to develop a rule for determining the perimeter of any number of regular polygons placed end-to-end.

COMMON CORE SMP SMP 3, SMP 7, SMP 8

MATERIALS Pouch: Pattern Blocks can be used for part of the activity.

GROUPING Work individually.

---

1. 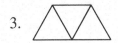   Perimeter = 3

2.    Perimeter = _____

3. Perimeter = _____

4. Perimeter = _____

5. What is the perimeter of 5 △s placed end-to-end as shown above? _____
    12 △s? _____

6. How many triangles would be needed to have a perimeter of 19? _____ 37? _____

7. Complete the following table.

Number of △s	1	2	3	4	5	9	13			28		100
Perimeter							19	23		35	

8. Given *k* triangles placed end-to-end, write a rule to determine the perimeter.

1. 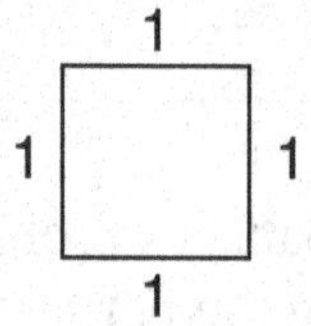 Perimeter = 4

2. 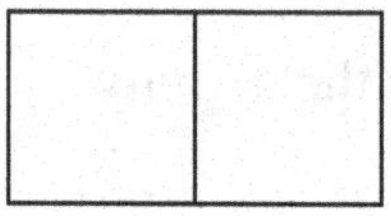 Perimeter = _____

3. 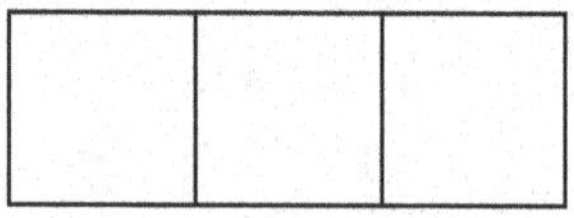 Perimeter = _____

4. What is the perimeter of 5 ⬜s placed end-to-end? _____ 9 ⬜s? _____

5. Complete the following table.

Number of ⬜s	1	2	3							100
Perimeter										

6. What is the perimeter of 13 ⬜s? _____ 23 ⬜s? _____

7. How many squares would be needed to have a perimeter of 38? _____ 66? _____

8. Can the perimeter of k squares be 45? Why or why not?

9. Given k squares placed end-to-end, write a rule to determine the perimeter.

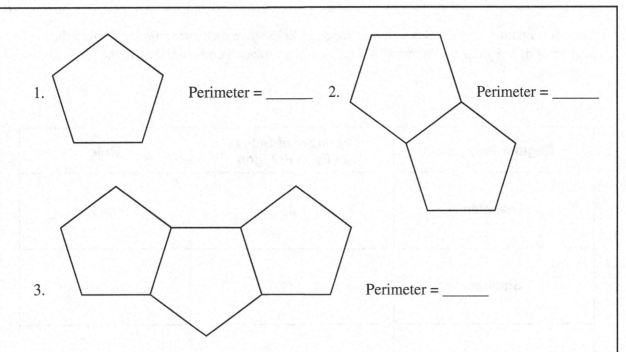

1.   Perimeter = _____ 2.   Perimeter = _____

3.   Perimeter = _____

4.   What is the perimeter of 8 ⬠s? _____ 17 ⬠s? _____

5.   How many pentagons would be needed to have a perimeter of 65? _____

6.   What is the perimeter of 100 ⬠s? _____ Explain how you arrived at your answer.

7.   Can the perimeter of *k* pentagons be 49? Why or why not?

8.   Given *k* pentagons placed end-to-end, write a rule to determine the perimeter.

---

1.   Given *k* regular hexagons placed end-to-end, write a rule to determine the perimeter.

2.   Given *k* regular heptagons (seven sides) placed end-to-end, write a rule to determine the perimeter.

3.   Given *k* regular octagons (eight sides) placed end-to-end, write a rule to determine the perimeter.

Look for a pattern in the rules you developed to help write a general rule for finding the perimeter of k regular polygons with n sides that are placed end-to-end as in the previous exercises.

Regular Polygons	Number of Sides in Each Polygon	Rule
Triangles	3	$k + 2$
Squares		
Pentagons		
Hexagons		
Heptagons		
Octagons		
⋮		
n-gons	n	

Activity 3: What's My Function?

PURPOSE	Determine the function relating input and output values.
COMMON CORE SMP	SMP 1, SMP 2, SMP 4, SMP 5
MATERIALS	Online: Function Cards – 1 and Function Cards – 2
GROUPING	Work in pairs.
GETTING STARTED	Players alternate turns trying to guess the function on one of the cards.

A player scores 5 points if the function is guessed on the first try, 3 points if it is guessed on the second try, and 1 point if it is guessed on the third try. Once the player has determined the function, 1 additional point is scored for each correct way the player can express the function: table, graph, equation, function, or verbal expression. The player who has scored the most points after all of the cards have been used is the winner.

To start the game, shuffle the cards and place them face down on the table. One player draws a card from the top of the deck without showing it to the other player. The other player tries to guess the function on the card by giving input values one at a time. The player holding the card applies the function on the card to the input value and responds with the output.

The *guesser* can give at most 10 input values and can guess the rule at any time. However, only 3 guesses are allowed per card.

Example:

What's My Function?
Multiply input by 2 and add 1.
$y = 2x + 1$ $f(x) = 2x + 1$

x	–1	0	1	2	3
y	–1	1	3	5	7

Input	Output	Guess	Response
2	5	Add 3 to the input.	Incorrect.
0	1	(No Guess)	
4	9	Multiply input by 2 and add 1. $y = 2x + 1$ $f(x) = 2x + 1$	Correct.

Score:	Guessed function on second try:	3 points
	Expressed function 3 ways	3 points
	Total	**6 points**

1. What strategies for choosing input values might make it easier to guess the function?

EXTENSION	Make up Function Cards of your own and use them to play the game.

Activity 4:   Exploring Linear Functions

PURPOSE   Use a geoboard to develop the concept of slope and explore the graphs of linear equations.

COMMON CORE SMP   SMP 1, SMP 2, SMP 3, SMP 4, SMP 5, SMP 8

MATERIALS   Online: Geoboards Dot Paper
Other: Geoboards and geo-bands
Note: Geoboards Dot Paper and a ruler may be used instead of geoboards and geo-bands.

GROUPING   Work individually or in pairs.

GETTING STARTED   A geoboard is a good physical model of the first quadrant of the Cartesian coordinate grid. Think of the lower left corner as the *origin*, the horizontal row of pegs on the bottom as the *x-axis*, and the vertical column of pegs on the left side as the *y-axis*.

1. a. Connect a geo-band from the origin to the peg in the top right corner. The geo-band represents a line on the geoboard.

    b. Complete the table below for the points (pegs) on the line.

x	0	1	2	3	4
y	0	1			

    c. How are the values of *x* and *y* related?

    d. Use your answer to Part c to write an equation for the line.

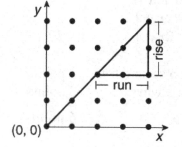

2. a. Connect any two pegs on the line with a geo-band. Pull one end of the geo-band down to form a right triangle as in the diagram at the left.

    b. Find the ratio of the *rise* to the *run* $\left(\dfrac{\text{rise}}{\text{run}}\right)$ for the two points you chose.

    c. Repeat Parts a and b for two different points on the line.

    d. What do you notice about the two ratios? Do you think this would be the same for any two points on the line? Why or why not?

The *slope* of a line is the ratio of the *rise* to the *run*. The slope is a measure of the steepness of the line.

3. a. Connect the origin and the point (4, 2) with a geo-band. Complete the table below for the points on the line.

x	0	1	2	3	4
y	0			$1\frac{1}{2}$	

   b. How are the values of *x* and *y* related?

   c. Use your answer to Part b to write an equation for the line.

   d. Find the slope of the line.

4. a. Connect the origin to the point (1, 4) on your geoboard with a geo-band. What is the slope of the line?

   b. How can you use the slope to find the value of *y* when $x = 2$?

   c. Use the slope to complete the table below for the points on the line.

x	0	1	2	3	4
y	0	4			

   d. Write an equation for the line.

5. a. Compare the slopes and the lines in Exercises 1 through 4. How does the slope of a line indicate how steep the line is?

   b. Compare the slopes and the equations of the lines in Exercises 1 through 4. How are the slopes of the lines related to the equations of the lines?

6. a. Connect the points $(0, 1)$ and $(4, 3)$ with a geo-band. What is the slope of the line?

   b. Complete the table below for the points on the line.

x	0	1	2	3	4
y	1				3

   c. What is the equation of the line?

   d. What is the y-coordinate of the point where the line intersects the y-axis? This is the *y-intercept*.

7. The lines in Exercises 1 through 4 went through the origin. What were their y-intercepts?

8. a. Connect the points $(0, 4)$ and $(2, 0)$ with a geo-band. What is the slope of the line?

   b. How does the slope indicate that the line goes downhill from left-to-right?

   c. What is the y-intercept?

   d. Complete the table below for the points on the line.

x	0	1	2	3	4
y	4		0		

   e. What is the equation of the line?

9. a. In Exercises 6 and 8, how are the slopes and the y-intercepts of the lines related to the equations of the lines?

   b. Is this also true for the lines in Exercises 1 through 4? Explain.

10. a. Construct a horizontal line on your geoboard. What is the slope of the line?

    b. What is the equation of the line?

11. a. Construct a vertical line on your geoboard. What is the slope of the line?

    b. What is the equation of the line?

12. a. Construct a pair of parallel lines on your geoboard.

    b. Find the slopes of the lines.

       slope of line 1: _____   slope of line 2: _____

    c. Construct two additional pairs of parallel lines on your geoboard. For each pair, find the slopes of the lines and record them below.

       Pair A: slope of line 1: _____   slope of line 2: _____

       Pair B: slope of line 1: _____   slope of line 2: _____

    d. Based on your results in Parts a–c, what conjecture can you make regarding the slopes of parallel lines?

13. a. Without using vertical and horizontal segments, construct a pair of perpendicular lines on your geoboard.

    b. Find the slopes of the lines.

       slope of line 1: _____   slope of line 2: _____

    c. Construct two additional pairs of perpendicular lines on your geoboard. For each pair, find the slopes of the lines and record them below.

       Pair A: slope of line 1: _____   slope of line 2: _____

       Pair B: slope of line 1: _____   slope of line 2: _____

    d. Based on your results in Parts a–c, what conjecture can you make regarding the slopes of perpendicular lines?

Activity 5: Graphing Rectangles

PURPOSE	Display discrete and continuous data using graphs and develop the idea of a limit. Reinforce the concepts of perimeter and area.
COMMON CORE SMP	SMP 1, SMP 2, SMP 3, SMP 4, SMP 5, SMP 6, SMP 7, SMP 8
MATERIALS	Pouch: 36 Colored Squares Online: Centimeter Graph Paper Other: Scissors and a loop of string 24 cm in circumference
GROUPING	Work individually or in pairs.

1. Determine all possible rectangles that can be constructed using the 36 squares. Record each rectangle on a piece of centimeter graph paper and label the base (*b*) and the height (*h*).

2. Measure the length of the base, the height, and the perimeter (*P*) for each rectangle and record the measurement in Table 1. The length of the side of a square equals 1 unit.

TABLE 1: Area = 36

Base (*b*)								
Height (*h*)								
Perimeter (*P*)								

3. Cut out each rectangle. Construct a graph showing only the first quadrant. Label the horizontal axis (*b*) and the vertical axis (*h*). Place each rectangle on the axis as illustrated below. Each rectangle must be placed so that one vertex is at (0, 0). Make a drawing of the completed graph.

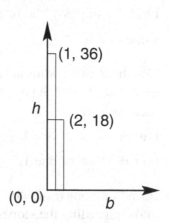

1. Using the data in Table 1, plot the ordered pairs (b, h) that represent the base and height of each rectangle on Graph 1.

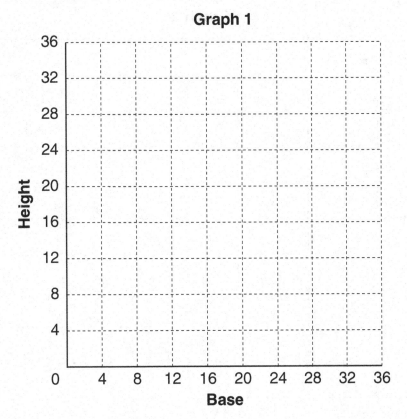

Graph 1

The data displayed in Graph 1 represent *all* rectangles that can be constructed using 36 squares.

2. Are there any other rectangles with an area of 36? Explain.

3. How many are there?

4. List the dimensions of at least three other rectangles that have an area of 36.

5. Are these rectangles represented in Graph 1? If not, describe where on the graph the new points should be placed.

6. Plot and connect the points that represent the ordered pairs (b, h) for **all** possible rectangles whose areas are 36.

7. Will the graph of this data intersect either axis? If so, where? If not, explain.

1. Using the data from Table 1, plot the ordered pairs (b, P) on Graph 2. Connect the points on the graph with a smooth curve.

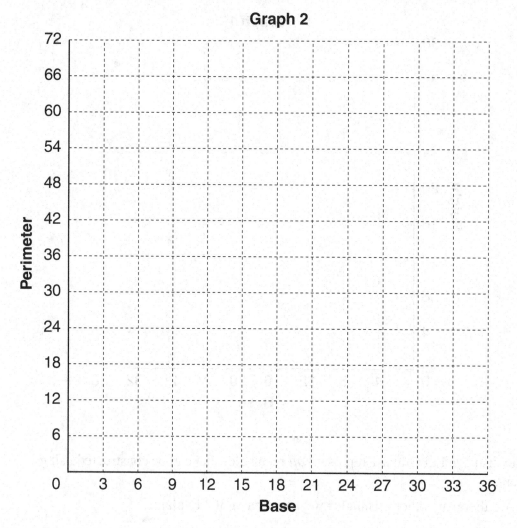

Graph 2

2. Each rectangle represented on the graph has an area of _____.

3. What is the least perimeter of any rectangle?

4. What are the dimensions of the rectangle with the least perimeter?

5. Is there a rectangle with a maximum perimeter? Explain.

6. Explain how it is possible for a rectangle to have an area of 36, yet have a perimeter that is unlimited.

7. Describe a physical model that you could use to illustrate the concept of a finite area being enclosed by an unlimited perimeter.

1. Determine all possible rectangles with integral dimensions that can be enclosed by the loop of string. Outline each rectangle on centimeter graph paper and label the base (*b*) and the height (*h*). What is the perimeter of each rectangle?

2. Record the base (*b*), the height (*h*), and the area (*A*) for each rectangle in Table 2.

TABLE 2: Perimeter = 24

Base (*b*)										
Height (*h*)										
Area (*A*)										

3. Using the data in Table 2, plot the ordered pairs (*b*, *h*) that represent the base and height of each rectangle on Graph 3. Draw a line through the points.

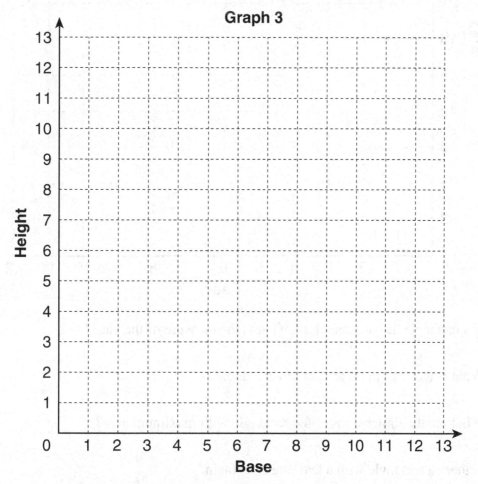

Graph 3

4. Graph 3 displays the data for **all** rectangles with a perimeter of 24 cm. Will the graph of this data intersect either axis? If so, where?

5. If the graph did intersect the horizontal axis, what would be the coordinates of that point? _____ How is the *b*-coordinate of that point related to the perimeter?

1. Using the data from Table 2, plot the ordered pairs (b, A) on Graph 4. Connect the points on the graph with a smooth curve.

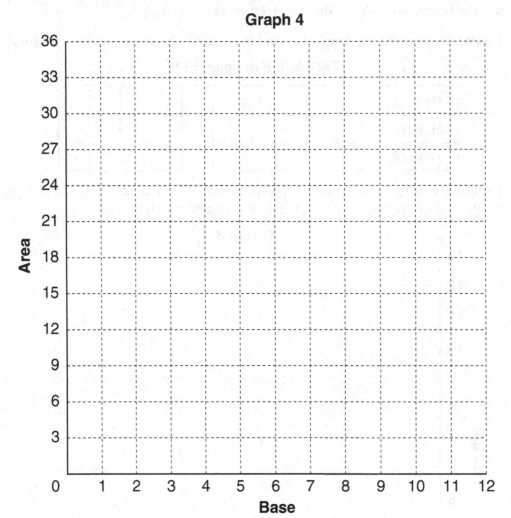

Graph 4

2. If a rectangle has an area of 24, what is the measure of the base?

3. What is the maximum area of any rectangle?

4. What are the dimensions of the rectangle with maximum area?

5. Is there a rectangle with a least area? Explain.

6. As the area of the rectangle approaches zero, the measure of the base approaches a maximum value of _____. Explain why this happens.

Chapter Summary

The goal of this chapter was to review some of the basic concepts of algebra and to develop algebraic reasoning. The concept of a function, a fundamental idea in mathematics was introduced through a variety of representations. Functions provide a powerful tool for modeling situations and connecting mathematical ideas.

Algebraic expressions were applied in geometric settings in Activities 1 and 2 where you discovered patterns and found rules to generalize the patterns. The activities reinforced the concept of perimeter and also helped to develop spatial perception.

Activity 3 built on the lessons learned in Chapter 1. The concept of a function was reinforced through a game that applied the guess-and-check strategy to input and output values in order to identify various functions. The game also emphasized multiple ways a function can be expressed: as a table, a graph, an equation, or using function notation.

In Activity 4, you used tables, graphs, and equations to develop the concept of slope and to explore the relationship of the slope and the y-intercept to the equation of a line.

Coordinate geometry was used in Activity 5 to study the relationship between area and perimeter. You discovered that, for a given area, there is a minimum perimeter that will enclose it, but no maximum perimeter. Thus it is possible to have a finite area bounded by an infinite perimeter. However, for a given perimeter, you found that it will enclose a maximum area, but no minimum area.

Chapter 9
Geometric Figures

"Through the study of geometry, students will learn about geometric shapes and structures and how to analyze their characteristics and relationships. Spatial visualization—building and manipulating mental representations of two- and three-dimensional objects and perceiving an object from different perspectives—is an important aspect of geometric thinking. Geometry is a natural place for the development of students' reasoning and justification skills, culminating in work with proof in the secondary grades. Geometric modeling and spatial reasoning offer ways to interpret and describe physical environments and can be important tools ... (for) solving problems in other areas of mathematics and in real-world situations ..."

—Principles and Standards for School Mathematics

"... (students) should develop clarity and precision in describing the properties of geometric objects and then classifying them by these properties into categories such as rectangle, triangle, pyramid, or prism. They can develop knowledge about how geometric shapes are related to one another and begin to articulate geometric arguments about the properties of these shapes."

—Principles and Standards for School Mathematics

"Using concrete models, drawings, and dynamic geometry software, students can engage actively with geometric ideas. With well-designed activities, appropriate tools, and teachers' support, students can make and explore conjectures about geometry and can learn to reason carefully about geometric ideas from the earliest years of schooling."

—Principles and Standards for School Mathematics

In his book, *A Mathematician's Delight*, the renowned mathematician, W. W. Sawyer wrote: "The best way to learn geometry is to follow the road which the human race originally followed:

> Do things, arrange things,
> Make things, and only then
> Reason about things."

The activities in this chapter follow Sawyer's road. They provide a variety of opportunities to arrange, measure, and construct geometric shapes in two and three dimensions before conjecturing and reasoning about the new figures, and then developing an understanding of their fundamental properties.

These problem-solving activities provide an opportunity to explore and discover geometric concepts and relationships and to investigate curves and explore the relationships among points that are inside or outside a simple closed curve.

Pattern blocks and geoboards will be used to explore angle measure and to discover the relationship between the number of sides of a convex polygon and the sum of the measures of its angles.

After constructing triangles with given properties, you will make and test conjectures based on your observations and comparison of your results with those of a classmate. Also, by constructing various two-dimensional nets and folding them into polyhedrons, you will strengthen your spatial visualization skills.

This informal exploration of shapes and their properties builds the foundation that is needed for the study of formal-deductive geometry.

Correlation of Chapter 9 Activities to the
Common Core Standards of Mathematical Practice

Activity Number and Title	Standards of Mathematical Practice
1: What's the Angle?	SMP 3, SMP 4, SMP 5, SMP 6, SMP 7, SMP 8
2: Triangle Properties—Angles	SMP 3, SMP 4, SMP 5, SMP 6, SMP 7, SMP 8
3: Inside or Outside?	SMP 2, SMP 4, SMP 5, SMP 6, SMP 7, SMP 8
4: Angles on Pattern Blocks	SMP 2, SMP 4, SMP 5, SMP 6, SMP 7, SMP 8
5: Sum of Interior/Exterior Angles	SMP 2, SMP 4, SMP 5, SMP 6, SMP 7
6: Stars and Angles	SMP 2, SMP 3, SMP, 5, SMP 6, SMP 7
7: Mysterious Midpoints	SMP 2, SMP 3, SMP 4, SMP 5, SMP 6, SMP 7, SMP 8
8: Spatial Visualization	SMP 2, SMP 3, SMP 4, SMP 5, SMP 6, SMP 7, SMP 8

Activity 1: What's the Angle?

PURPOSE   Develop angle measurement using the central angle in a circle and its intercepted arc.

COMMON CORE SMP   SMP 3, SMP 4, SMP 5, SMP 6, SMP 7, SMP 8

GROUPING   Work individually.

GETTING STARTED   The measure of an angle is the amount of rotation as a ray turns from coinciding with one side of the angle to coinciding with the other side. A complete rotation of a ray about a point results in an angle with a measure of **360°**. One degree is $\frac{1}{360}$ of a complete rotation.

A **central angle** is an angle whose vertex is the center of a circle and whose sides contain radii.

Example:

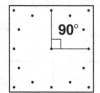

For each central angle, record its measure and its classification: *acute*, *right*, or *obtuse*.

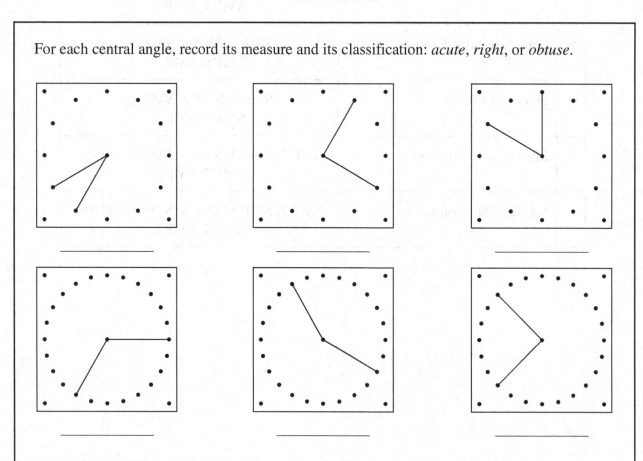

For each exercise, construct an angle that has the given measurement. In some exercises, one side of the angle is given.

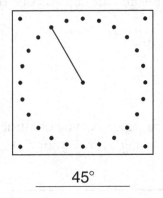

_____
45°

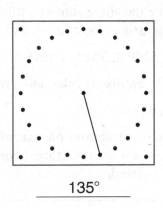

_____
135°

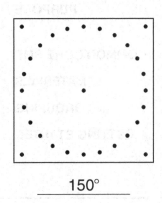

_____
150°

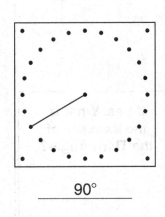

_____
90°

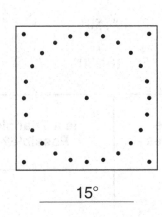

_____
15°

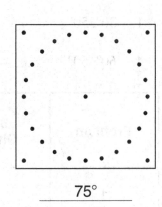

_____
75°

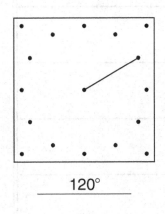

_____
120°

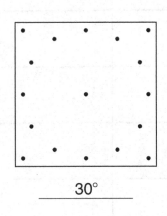

_____
30°

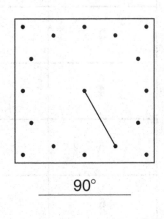

_____
90°

Activity 2: Triangle Properties—Angles

PURPOSE	Reinforce the angle sum of a triangle theorem and the construction of a triangle using a protractor and ruler.
COMMON CORE SMP	SMP 3, SMP 4, SMP 5, SMP 6, SMP 7, SMP 8
MATERIALS	Other: A centimeter ruler and a protractor
GROUPING	Work in pairs.
GETTING STARTED	Make all constructions on a separate piece of paper. As you construct the triangles in each section, compare your triangles with your partner's triangles.

Use a ruler and a protractor to construct a triangle that has two angles with the indicated measures. Record your results in the table.

1. 30°, 50°

2. 40°, 50°

3. 90°, 95°

4. 60°, 60°

5. 110°, 70°

6. 80°, 80°

Problem	Sum of the Given Angles	Is a Triangle Possible?	If Yes, What Is the Measure of the Third Angle?
1			
2			
3			
4			
5			
6			

1. In each Exercise 1–6 that **did** result in a triangle, what is true about the sum of the measures of the two given angles?

2. In each Exercise 1–6 that **did not** result in a triangle, what is true about the sum of the measures of the two given angles?

3. List the measures for three pairs of angles that **can** be used to construct a triangle.

   a. _____   b. _____   c. _____

4. List the measures for three pairs of angles that **cannot** be used to construct a triangle.

   a. _____   b. _____   c. _____

5. What can you conclude about the sum of the measures of the three angles of a triangle?

6. a. For each exercise that resulted in a triangle, list the angle pair and identify the type of triangle that was constructed.

   b. What did you observe when you compared your triangle to your partner's triangle?

Activity 3: Inside or Outside?

PURPOSE	Explore the relationships among points in the interior and exterior of a simple closed curve.
COMMON CORE SMP	SMP 2, SMP 4, SMP 5, SMP 6, SMP 7, SMP 8
GROUPING	Work individually.
GETTING STARTED	The following examples illustrate various curves.

Examples:

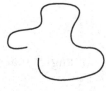

Exterior

Simple
closed

Simple
not closed

Not Simple
closed

Some *simple* curves may not appear to be so simple. In the curve shown below, point K is clearly outside the curve.

1. Do you think point M is inside the curve or outside?

2. Do you think point D is inside the curve or outside?

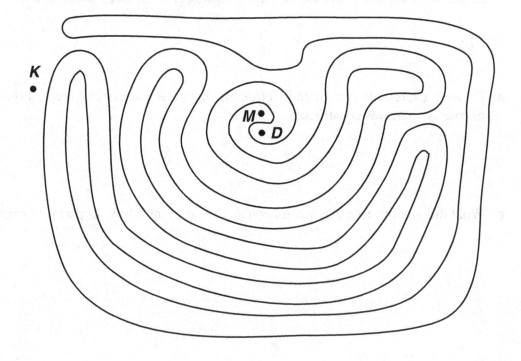

For each of the following curves, point *M* and point *B* are either both inside or both outside the curve. Without crossing a boundary, draw a curve from *M* to *B*. Then draw a segment from *M* to *B* and count the number of times the segment crosses the boundary. (A *crossing* means that the segment goes from inside the curve to the outside, or from outside the curve to the inside.) Record the number of times the segment crosses the boundary in the table below.

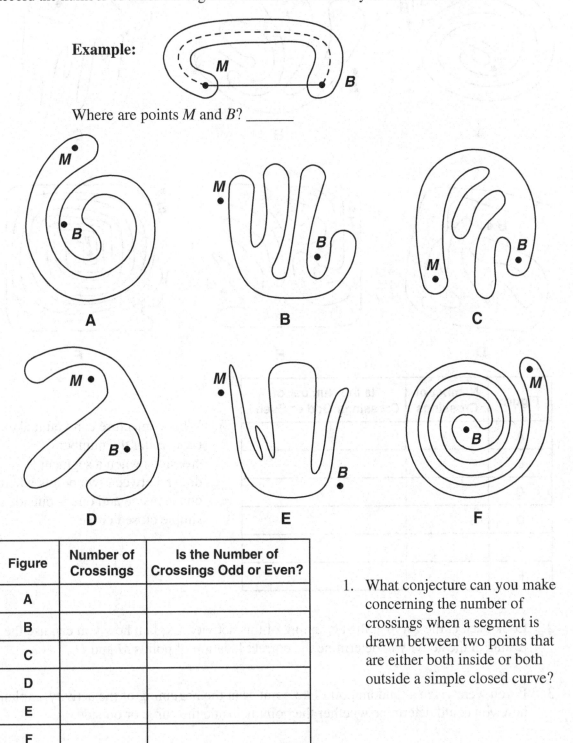

Example:

Where are points *M* and *B*? _____

A

B

C

D

E

F

Figure	Number of Crossings	Is the Number of Crossings Odd or Even?
A		
B		
C		
D		
E		
F		

1. What conjecture can you make concerning the number of crossings when a segment is drawn between two points that are either both inside or both outside a simple closed curve?

In each of the curves, point *T* is inside the curve and point *B* is outside. Draw a segment from *T* to *B*, count the number of times the segment crosses the boundary, and complete the table.

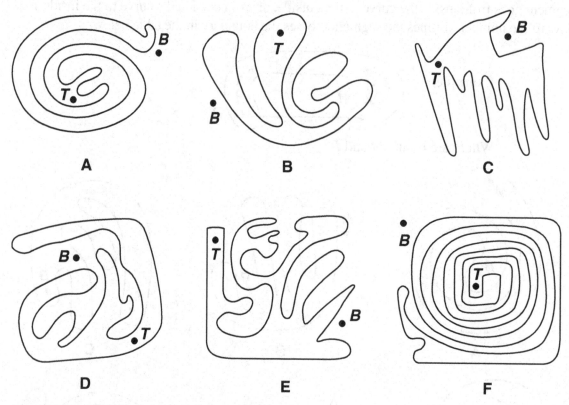

A   B   C

D   E   F

Figure	Number of Crossings	Is the Number of Crossings Odd or Even?
A		
B		
C		
D		
E		
F		

1. What conjecture can you make concerning the number of crossings when a segment is drawn between two points where one is inside and one is outside a simple closed curve?

2. Look back at the curve at the beginning of this activity. Explain how you can use the results of the activity to determine the correct location of points *M* and *D*.

3. If you were given a random point like point *M* at the beginning of the activity, explain how you could determine whether the point is inside the curve or outside.

Activity 4: Angles on Pattern Blocks

PURPOSE	Determine the sum of the measures of the interior angles of a polygon and the measure of each angle of a regular polygon.
COMMON CORE SMP	SMP 2, SMP 4, SMP 5, SMP 6, SMP 7, SMP 8
MATERIALS	Pouch: Pattern Blocks
GROUPING	Work individually.

Determine the measure of each interior angle of each pattern block. You may use **only** the fact that the square has four right angles. **HINT:** You may place combinations of blocks on top of a block to assist in determining the measures of the angles.

Example: Indicate the measure of each angle inside the pattern blocks as shown in the example. For each pattern block, explain the method you used to determine the measure of each angle. Draw a sketch of the blocks you used to illustrate your explanations. The measures you find on one block may be used to determine the measures of the angles on other blocks.

1.

2.

3.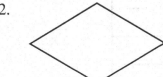

4.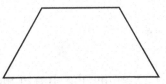

5.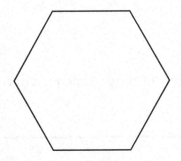

1. Use pattern blocks to construct a convex pentagon. Determine the measure of each interior angle of the pentagon. Sketch your pentagon and indicate the measure of each angle.

2. Use pattern blocks to construct a convex heptagon (seven sides). Determine the measure of each interior angle of the heptagon. Sketch your heptagon and indicate the measure of each angle.

3. Use the measures of each of the interior angles of the pattern blocks and the polygons you constructed to complete the following table.

Number of Sides	Sum of the Measures of the Angles
3	
4	
5	
6	
7	
8	
9	
n	

4. How is the increase in the **sum of the measures** of the interior angles related to the increase in the number of sides?

5. Given a convex polygon with n sides, how would you determine the **sum** of the measures of the interior angles of the polygon?

6. If a polygon is **regular**, how would you determine the measure of **each** interior angle?

Activity 5: Sum of Interior/Exterior Angles

PURPOSE Develop the relationships between the number of sides of a convex polygon and the sums of the measures of the interior and the exterior angles.

COMMON CORE SMP SMP 2, SMP 4, SMP 5, SMP 6, SMP 7

MATERIALS Other: The Geometer's Sketchpad® or other geometry drawing software

GROUPING Work individually.

Use the geometry software to construct a convex polygon with five sides. Measure each interior angle and find the sum of the measures. Using the *drag* feature, pick one vertex and move it around to alter the figure.

Observe any changes in the measures of each angle and the **sum** of the measures of the interior angles. For one polygon, record the measure of each angle and the sum of the measures of the interior angles in the table.

Repeat the process for polygons with 6, 7, 8, and 9 sides.

Polygon	Measure of Each Interior Angle									Sum
	1	2	3	4	5	6	7	8	9	
5 sides										
6 sides										
7 sides										
8 sides										
9 sides										

1. What is the difference between successive sums in the table? _____

2. Write a rule to determine the sum of the measures of interior angles of a convex polygon given the number of sides (*n*).

Use the geometry software to construct a polygon with five sides. Then, beginning at one vertex and continuing clockwise, extend each side of the polygon to form one *exterior angle* at each vertex as in the figure below.

Example:

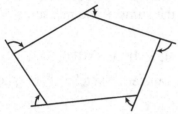

Measure each exterior angle, and determine the sum of the measures. Using the *drag* feature, grab one vertex and move it around to alter the figure. Observe any changes in the measure of each exterior angle and the **sum** of the measures of the exterior angles. Record the measure of each exterior angle and the sum of the measures of the exterior angles for one polygon in the table. Repeat the process for polygons with 6, 7, 8, and 9 sides.

Polygon	\multicolumn{9}{c}{Measure of Each Exterior Angle}	Sum
	1	2	3	4	5	6	7	8	9	
5 sides										
6 sides										
7 sides										
8 sides										
9 sides										

1. What can you conclude about the sum of the measures of the exterior angles of a convex polygon? Explain.

EXTENSION
1. What is the sum of the measures of an interior angle of a convex polygon and one of its adjacent exterior angles? _____

2. Complete the following for a convex polygon with (n) sides.

   a. The sum of the measures of the interior angles and one adjacent exterior angle at each vertex is _____.

   b. The sum of the measures of the interior angles is _____.

3. Use the results in Exercise 2 to prove algebraically that the sum of the measures of the exterior angles of a convex polygon, one at each vertex, is 360°.

Activity 6: Stars and Angles

PURPOSE   Discover a relationship between the number of points on a star and the measures of its interior and exterior angles.

COMMON CORE SMP   SMP 2, SMP 3, SMP 5, SMP 6, SMP 7

MATERIALS   Other: A protractor, ruler, and compass

GROUPING   Work individually or with a partner.

GETTING STARTED   The shape of a star is determined by the measure of the interior angle at each point and the measure of an exterior angle formed by adjacent arms. Different shape stars can be constructed by using different combinations of angle measures.

Examples:

The measure of the interior angle at each point of this four-pointed star is 40° and the measure of the exterior angle between adjacent arms is 130°.

The measure of the interior angle at each point of this five-pointed star is 36° and the measure of the exterior angle is 108°.

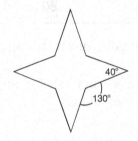

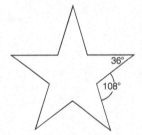

1. a. On a separate sheet of paper, construct the stars with the angle measures shown in the table.

Measure of the interior angle at each point (a)	Measure of the exterior angles between adjacent arms (b)	b – a	Number of points (N)
30°	150°	120°	
20°	92°		
20°	80°		
15°	60°		8
10°	46°		

b. Complete the last two columns in the table for the stars you constructed in Part a.

2. a. Look for a pattern in the table on the preceding page. How does the number of points on a star appear to be related to $b - a$?

    b. Write an equation that shows how N is related to $b - a$.

    c. Does your equation work for the stars in the Examples? Explain.

3. a. Predict whether each combination of angles in the following table will make a star. If a combination will make a star, predict the number of points it will have.

Measure of the interior angle at each point (a)	Measure of the exterior angles between adjacent arms (b)	Makes a star? (yes/no)	Number of points (N)
33°	105°		
25°	70°		
20°	100°		
15°	60°		

    b. Check your predictions by trying to construct each star.

4. Give 3 different combinations for angle measures a and b that will form a nine-pointed star.

EXTENSION

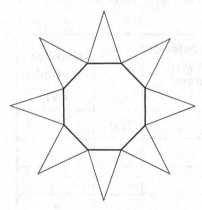

1. a. What type of polygon is formed by drawing segments connecting the non-point vertices of an N-pointed star as in the example at the left?

    b. What is the measure of an interior angle of this polygon?

    c. What type of triangle is formed at each point?

    d. What is the measure of each of the base angles in the triangle?

    e. What is the sum of the four angles formed at each non-point vertex of the star?

    f. Use your results from Parts b, d, and e to show that the equation you wrote in Exercise 2b is true for all stars.

Activity 7: Mysterious Midpoints

PURPOSE   Apply coordinate geometry techniques to reinforce understanding of the properties of quadrilaterals.

COMMON CORE SMP   SMP 2, SMP 3, SMP 4, SMP 5, SMP 6, SMP 7, SMP 8

MATERIALS   Online: Centimeter Graph Paper
Other: The Geometer's Sketchpad® or other geometry drawing software and a ruler

GROUPING   Work individually or in small groups.

---

1. Plot and label each of the following sets of points on a separate pair of coordinate axes. Draw four quadrilaterals by drawing segments connecting the points of each set in order.

    a. $P(2, 5)$, $I(7, 2)$, $N(12, 5)$, $K(7, 8)$ b. $B(-1, -5)$, $R(4, 2)$, $O(-3, 7)$, $W(-8, 0)$

    c. $R(2, 2)$, $O(-3, -1)$, $S(-6, -5)$, $E(4, -2)$ d. $P(9, 4)$, $O(11, 8)$, $L(2, 4)$, $Y(0, 0)$

2. Mark the midpoints of the sides of each quadrilateral and label them consecutively M, A, T, and H. Connect the points in order to form another quadrilateral. What appears to be true about each of the polygons *MATH*?

3. Explain how you can use coordinate methods to check your conjecture.

---

1. Plot and label the following sets of points on separate coordinate axes. Connect the points in each set in order to form the following special quadrilaterals: a parallelogram, a rhombus, a rectangle, and a square.

    a. $D(4, 8)$, $U(1, 3)$, $C(10, 6)$, $K(13, 11)$ b. $B(2, -2)$, $I(9, 1)$, $K(2, 4)$, $E(-5, 1)$

    c. $D(2, 3)$, $A(-3, 3)$, $V(-3, -4)$, $E(2, -4)$ d. $P(-2, -5)$, $I(-7, 0)$, $C(-2, 5)$, $K(3, 0)$

2. Identify each of the quadrilaterals and explain how you determined your answer.

3. Locate the midpoints of the sides and label them consecutively M, O, N, and T. Connect the points in order to form quadrilaterals.

    a. Are any of the quadrilaterals *MONT* that were formed special quadrilaterals? Explain how you determined your answer.

    b. Are any of the quadrilaterals *MONT* the same type of quadrilateral as the original figure? Explain how you determined your answer.

4. In each of the quadrilaterals *MONT* in Exercise 3, locate the midpoints of the sides, label them consecutively W, X, Y, and Z, and connect them in order to form new quadrilaterals. How are the new quadrilaterals related to the original figures drawn in Exercise 1?

1.   Complete each of the following statements.

   a. If you connect the midpoints of the sides of a quadrilateral in order, the resulting figure is a _____.

   b. If you connect the midpoints of the sides of a parallelogram in order, the resulting figure is a _____.

   c. If you connect the midpoints of the sides of a rectangle in order, the resulting figure is a_____.

   d. If you connect the midpoints of the sides of a rhombus in order, the resulting figure is a _____.

   e. If you connect the midpoints of the sides of a square in order, the resulting figure is a _____.

2.   Explain which properties of the original quadrilateral determine the special quadrilateral that is formed by connecting the midpoints of the sides.

EXTENSION

1.   Use a geometry drawing program to construct a quadrilateral. Locate the midpoints of the four sides and connect them in order as you did to form the polygon *MATH*. Use the dynamic feature of the program to alter the quadrilateral in various ways to verify that your first conjecture was correct.

2.   On separate screens, construct a trapezoid, parallelogram, rhombus, rectangle, kite, and square. In each figure, locate the midpoints of the sides and connect them in order. Then locate the midpoints of the sides of the new figure and connect them in order to form a second quadrilateral. Use the dynamic feature of the program to alter each of the original figures in various ways.

   Given that you start with a special quadrilateral, explain the relationship between the original quadrilateral and the quadrilateral that is formed by connecting the midpoints, and between the original quadrilateral and the second quadrilateral formed. Check your conjectures using the measuring capabilities of the software.

Activity 8: Spatial Visualization

PURPOSE	Develop spatial visualization through exploration of polyhedrons and their corresponding polyhedral nets.
COMMON CORE SMP	SMP 2, SMP 3, SMP 4, SMP 5, SMP 6, SMP 7, SMP 8
MATERIALS	Online: Centimeter Graph Paper, 3 copies of Nets for the Open-Top Box, and Polyhedral Nets (*Polydrons*™ may be used instead) Other: Scissors, tape, and colored markers
GROUPING	Work in pairs or individually.

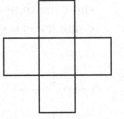

Cut out one net for an open-top box, and fold it to form a box.

1. Unfold the net, cut off one square. Move it to a new position and tape it to the adjacent square to form a new net which will also fold into an open-top box. Record your answer on graph paper. Find all possible solutions.

2. Cut out another net for an open-top box and cut off any two squares. Move the squares to new positions; tape them to adjacent squares so that the new net will fold into an open-top box. (A net may **not** be congruent to a net formed in Exercise 1.) Record your answer on graph paper. Find all possible solutions.

When you complete Exercises 1 and 2, you should have seven new nets, no two of which are congruent, and all of which will fold into an open-top box.

Cut out eight nets, fold each into an open-top box, and tape the edges to form eight boxes.

3. Pick one open-top box and one of the eight nets from Exercises 1 and 2. What edges should you cut along so that the box folds open to match the chosen net? Plan carefully. Mark the edges you plan to cut with a colored marker, then cut along the marked edges and unfold the box. If the unfolded box does not match the net you chose, refold the box, tape the edges, and try again.

4. Repeat the process for the remaining seven nets.

Draw the net shown at the right on graph paper. Cut it out and fold it to form a cube.

1. Cut off the square marked x, move it to a new position, and tape it to an adjacent square to form a new net that will also fold into a cube. Record your answer on graph paper. Find all possible solutions so that no two nets are congruent.

2. When the net at the right is folded into a cube, what faces are opposite each other? Write three equations to show the sums of the pairs of numbers on opposite faces.

____ + ____ = ____   ____ + ____ = ____   ____ + ____ = ____

3. Write the missing digits on each net so that when you cut it out and fold it into a cube, the sums of the opposite faces of the cube are the same as those in Exercise 2.

4. Desk calendars are made with number cubes where each face of a cube has one of the digits 0 through 9 on it. The cubes can be arranged so the faces will show the dates 01, 02, … 31. What are the missing digits on each cube?

Right cube   ____, ____, ____
Left cube   ____, ____, ____, ____

Each net in the left column will fold into a polyhedron. Use **Y** or **N** to indicate which of the nets to the right will also fold into the same polyhedron. If necessary, cut out the Polyhedral Nets or make the nets with *Polydrons*™. Then check your conjecture by folding the nets into polyhedra.

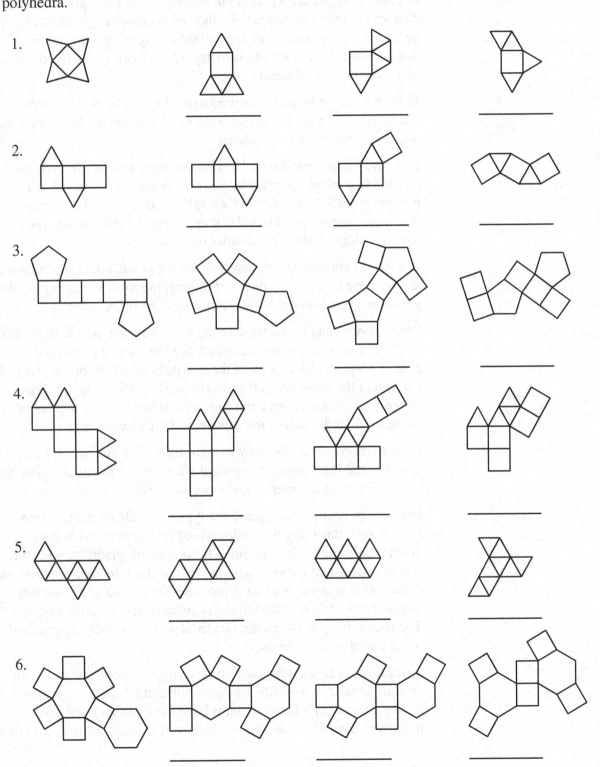

Chapter Summary

Geometric shapes are a part of our everyday life. Throughout this chapter, the use of manipulatives allowed hands-on exploration of geometric figures and promoted understanding of the basic concepts that are used in the study of geometry. The activities introduced you to several important geometric ideas.

In Activity 1, geoboards were used to explore angle measurement. Classifying the angles as *acute*, *right*, or *obtuse* laid the foundation for later exercises involving triangles.

In Activity 2, you were introduced to the angle sum of a triangle through construction using a protractor and ruler. When you constructed a triangle given the measures of the angles, and compared your triangle with your partner's, you found that the triangles were similar. They had the same shape, but not necessarily the same size.

Activity 3 explored the relationships among points inside and outside a simple closed curve. You discovered a method for determining whether a random point is in the *interior* or the *exterior* of the curve.

Pattern blocks and geometry drawing software were used in Activities 4 and 5 to develop the relationship between the number of sides of a convex polygon and the sum of the measures of the interior angles and the sum of the measures of the exterior angles. After using inductive reasoning to make a conjecture, you were asked to prove algebraically a theorem about the sum of the measures of the exterior angles.

In Activity 6, you extended what you learned in Activities 4 and 5 to discover the relationship between the number of points on a star and the measures of its interior and exterior angles.

In Activity 7, you investigated the type of quadrilateral that was formed by connecting the midpoints of the consecutive sides of a given quadrilateral. You found that the type of quadrilateral formed was dependent on the original one. Connecting the midpoints of the consecutive sides of the new quadrilateral resulted in yet another quadrilateral, again related to the previous one and to the original. The relationship between the quadrilaterals was easily determined using coordinate methods.

Several approaches were used in Activity 8 to help enhance your spatial visualization skills. First, you investigated the arrangements of five squares in a net that could be folded into an open-top box—a transformation from two-dimensions to three-dimensions. You then had to cut along the edges of the box so that it would unfold into one of the nets—a transformation from three-dimensions to two-dimensions. Lastly, several different nets were given and you had to guess which ones would fold into the same polyhedron. You checked the accuracy of your predictions by constructing each net and attempting to fold it into the given polyhedron.

Chapter 10
Measurement: Length, Area, and Volume

"The study of measurement is important in the mathematics curriculum from prekindergarten through high school because of the practicality and pervasiveness of measurement in so many aspects of everyday life. The study of measurement also offers an opportunity for learning and applying other mathematics, including number operations, geometric ideas, statistical concepts, and notions of function. It highlights connections within mathematics and between mathematics and areas outside of mathematics. ..."

—*Principles and Standards for School Mathematics*

"Students understand the statement of the Pythagorean Theorem and its converse, and can explain why the Pythagorean Theorem is valid, for example, by decomposing a square in two different ways. They apply the Pythagorean Theorem to find distances between points on a coordinate plane, to find lengths, and to analyze polygons."

—*Common Core State Standards for Mathematics*

The activities in this chapter follow W.W. Sawyer's recommendation in *A Mathematician's Delight* to do things, to arrange things, and to make things before reasoning about them. Through exploration with a variety of manipulatives, you will develop an understanding of the concepts of area and volume and the formulas for finding the areas of certain polygons and the volumes of prisms, pyramids, cylinders, and cones.

All of the formulas for area will be developed sequentially beginning with the relationship of the product of the dimensions of a rectangular array used to illustrate multiplication. Each new formula will be related to a previously developed one by comparing the actual areas of the two polygons. A similar sequence of activities will develop the rules for determining the volumes of certain polyhedra.

The Pythagorean theorem, one of the most important theorems of geometry, will be explored informally through two puzzles. Each one involves manipulatives to clearly demonstrate the relationship between the area of the square constructed on the hypotenuse and the sum of the areas of the squares constructed on the other two sides.

175

Correlation of Chapter 10 Activities to the
Common Core Standards of Mathematical Practice

Activity Number and Title		Standards of Mathematical Practice
1:	Areas of Polygons	SMP 3, SMP 4, SMP 5, SMP 7
2:	From Rectangles to Parallelograms	SMP 3, SMP 4, SMP 5, SMP 7
3:	From Parallelograms to Triangles	SMP 3, SMP 4, SMP 5, SMP 7
4:	From Parallelograms to Trapezoids	SMP 3, SMP 4, SMP 5, SMP 7
5:	Pythagorean Puzzles	SMP 3, SMP 4, SMP 5, SMP 7
6:	Now You See It, Now You Don't	SMP 2, SMP 3, SMP 4, SMP 5, SMP 7
7:	Right or Not?	SMP 2, SMP 3, SMP 4, SMP 5, SMP 7
8:	Volume of a Rectangular Solid	SMP 2, SMP 3, SMP 4, SMP 5, SMP 6, SMP 7
9:	Pyramids and Cones	SMP 3, SMP 4, SMP 5, SMP 6, SMP 7
10:	Surface Area	SMP 3, SMP 4, SMP 5, SMP 6, SMP 7

Activity 1: Areas of Polygons

PURPOSE   Develop the concept of the area of a polygon through the composition and decomposition of non-overlapping parts of a figure and reinforce the concept of convex and concave polygons.

COMMON CORE SMP   SMP 3, SMP 4, SMP 5, SMP 7

MATERIALS   Online: Dot Paper for recording
Other: Geoboard and geo-bands

GROUPING   Work individually.

GETTING STARTED   The smallest square that can be constructed on your geoboard has an area of one square unit.

---

1. On your geoboard, construct **10 or more** polygons that have an area of $2\frac{1}{2}$ square units and record your results on dot paper. Carefully check to be sure that each new figure is not congruent to a previous figure.

2. Sort the figures into sets of concave and convex polygons.

---

1. On your geoboard, construct **10 or more** polygons with an area of 3 square units and **10 or more** polygons with an area of $3\frac{1}{2}$ square units. Record your results on dot paper. Carefully check to see that each new polygon is not congruent to a previous figure.

2. Sort the figures into sets of concave and convex polygons.

3. Look at the figures that you constructed with an area of 3 square units. Are the perimeters of all of the polygons equal? **NOTE:** The distance between two adjacent horizontal or vertical pegs is one unit. The distance between any other pair of pegs is more than one unit.

   a. What is the least perimeter?

   b. What is the greatest perimeter (approximately)?

   c. If you had a 1000 × 1000 peg geoboard, could you construct a polygon with an area of 3 square units and a perimeter of 100? 1000? Explain.

Activity 2:   From Rectangles to Parallelograms

PURPOSE	Develop a formula for finding the area of a parallelogram by comparing the area of a parallelogram to the area of a related rectangle.
COMMON CORE SMP	SMP 3, SMP 4, SMP 5, SMP 7
MATERIALS	Online: Dot Paper Other: Ruler and scissors
GROUPING	Work individually.

Construct a parallelogram on your dot paper. Construct the altitude from one vertex of the upper base as shown. Cut out the parallelogram and then cut off the triangle. Move the triangle to the other end of the figure and match the vertices as shown.

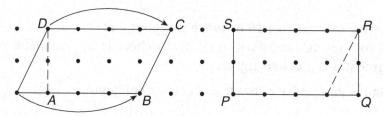

1. What kind of polygon is the new figure?

2. What are the properties of polygon *PQRS* that support your answer in Question 1.

3. What is the relationship between the base and the altitude of the original parallelogram and those of the new polygon?

4. What is the area of the new polygon?

5. What is the relationship between the area of the original parallelogram and the area of the new polygon?

6. Describe two methods for finding the area of the new polygon.

Construct five additional parallelograms on your dot paper. Draw an altitude, cut out the figures, and construct a rectangle as shown above.

1. For each new parallelogram, determine the area of the related rectangle.

2. What is the relationship between the area of the parallelogram and the area of the related rectangle?

3. Write a rule for determining the area of a parallelogram.

Activity 3: From Parallelograms to Triangles

PURPOSE	Develop a formula for finding the area of a triangle by comparing the area of a triangle to the area of a related parallelogram.
COMMON CORE SMP	SMP 3, SMP 4, SMP 5, SMP 7
MATERIALS	Online: Dot Paper Other: Ruler
GROUPING	Work individually.

Construct $\triangle KIM$ on your dot paper. Construct a segment $\overline{KE}$ that is parallel to $\overline{MI}$ and has length equal to the length of $\overline{MI}$ as shown. Draw the segment $\overline{EM}$.

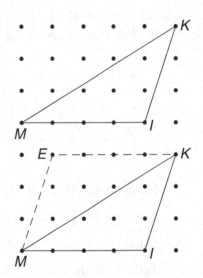

1. Polygon *MIKE* is what type of quadrilateral?

2. What are the properties of polygon *MIKE* that support your answer in Question 1.

3. What is the area of quadrilateral *MIKE*?

4. The area of $\triangle KIM$ is what fractional part of the area of the quadrilateral *MIKE*?

5. What is the area of $\triangle KIM$?

Construct five additional triangles on your dot paper. Then construct the related parallelogram as shown above.

1. For each new triangle, what is the relationship between the area of the triangle and the area of the related parallelogram?

2. For each new triangle, what is the relationship between the base and altitude of the triangle and those of the related parallelogram?

3. What is the formula for finding the area of a parallelogram?

4. Use the relationship between the area of a triangle and the area of a related parallelogram to write a rule for determining the area of a triangle.

Activity 4: From Parallelograms to Trapezoids

PURPOSE	Develop a formula for finding the area of a trapezoid by comparing the area of a trapezoid to the area of a related parallelogram.
COMMON CORE SMP	SMP 3, SMP 4, SMP 5, SMP 7
MATERIALS	Online: Dot Paper Other: Ruler
GROUPING	Work individually.

Construct a trapezoid *ABCD* on your dot paper as shown. Construct a rotated copy of trapezoid *ABCD*, and match the vertices with the original trapezoid as shown.

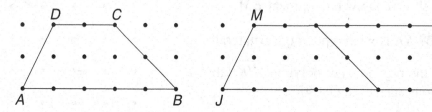

1. Polygon *JKLM* is what type of quadrilateral?

2. What are the properties of polygon *JKLM* that support your answer in Question 1.

3. What is the area of quadrilateral *JKLM*?

4. What is the relationship between the area of trapezoid *ABCD* and the area of quadrilateral *JKLM*?

5. What is the area of trapezoid *ABCD*?

Construct five additional trapezoids on your dot paper. Then construct the related parallelogram.

1. For each trapezoid, what is the relationship between the area of the trapezoid and the area of the related parallelogram?

2. What is the relationship between the measure of the base of the parallelogram and the measures of the two bases of the trapezoid?

3. Use the relationship between the area of a trapezoid and the area of a related parallelogram to write a rule for determining the area of a trapezoid.

Activity 5: Pythagorean Puzzles

PURPOSE	Use puzzles to explore the Pythagorean theorem.
COMMON CORE SMP	SMP 3, SMP 4, SMP 5, SMP 7
MATERIALS	Online: Pythagorean Puzzles sheet Other: Scissors and protractor
GROUPING	Work individually.

PUZZLE 1

1. Measure the three angles of $\triangle XYZ$ in Puzzle 1.

2. $\triangle XYZ$ is a _____ triangle.

3. Cut out Square A and the four pieces of Square B. Put them together to form Square C.

   What conclusion can you make about the sum of the areas of Square A and Square B as compared to the area of Square C?

PUZZLE 2

- Cut out the five puzzle pieces in Puzzle 2.
- Put the two triangles and the two pentagons together to form a square.
- Place this square on a sheet of paper. Trace around the perimeter; cut out the square and label it Square B.
- Now, put the original five pieces together to form a **larger** square.
- Place this square on a sheet of paper. Trace around the perimeter; cut out the square and label it Square C.

1. What conclusion can you make about the sum of the areas of Square A and Square B as compared to the area of Square C?

2. Place Square A, Square B, and Square C together as in Puzzle 1 to form a triangle.

   The triangle formed is a _____ triangle.

Activity 6: Now You See It, Now You Don't

PURPOSE	Explore the concept of conservation of area in a problem-solving setting.
COMMON CORE SMP	SMP 2, SMP 3, SMP 4, SMP 5, SMP 7
MATERIALS	Online: Centimeter Graph Paper Other: Ruler, scissors, and calculator
GROUPING	Work individually.

1. On a piece of graph paper, draw an 8×8 square and divide it into four parts as shown in the figure at the right. What is the area of the square?

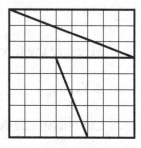

2. Cut out the four parts and form a rectangle as shown. What is the area of the rectangle? Can you explain this?

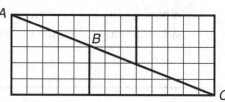

3. Rearrange the four parts as shown. What is the area of this figure? Is this possible?

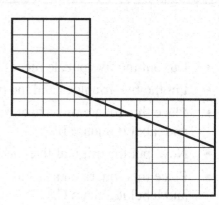

4. Use the Pythagorean theorem to compare $AB + BC$ to AC. What did you find? What does it mean?

The Fibonacci numbers are found in a famous sequence of numbers: 1, 1, 2, 3, 5, 8, . . . Each number in the sequence is the sum of the previous two numbers.

What are the next four Fibonacci numbers?

Note that the length of a side of the square was a Fibonacci number and that three sides were divided into two parts. The measure of each part was also a Fibonacci number.

1. On graph paper, draw a 13 × 13 square. Divide the square into four parts as shown. In this case, divide three consecutive sides into two parts with measures 5 and 8. Note that, as in the previous exercise, the length of the side of the square (13) and the lengths of the two parts (5 and 8) are Fibonacci numbers.

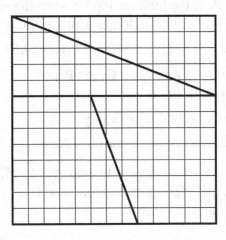

2. What is the area of the square?

3. Rearrange the four parts into a rectangle as in Exercise 2 on the previous page. What is the area of the rectangle?

4. Rearrange the four parts into the figure shown in Exercise 3 on the previous page. What is the area of this shape?

---

1. What is the next Fibonacci number after 13?

2. If you were to construct a square with that number as the length of the side, how would you divide the sides in order to make the four parts as in the previous exercises?

3. If you rearrange the four parts to form a rectangle and the other shape as before, what would be the area of the rectangle? _____ of the other shape? _____

EXTENSION   For any number in the Fibonacci sequence ($n \geq 3$), what is the relationship between the area of the square with that number as the length of a side and the areas of the rectangle and the other shape that can be constructed if the original square is divided and rearranged twice as was done in this activity?

Activity 7: Right or Not?

PURPOSE	Develop or reinforce the Pythagorean theorem and its converse and explore the relationship between the lengths of the sides of a triangle and its classification as acute, right, or obtuse.
COMMON CORE SMP	SMP 2, SMP 3, SMP 4, SMP 5, SMP 7
MATERIALS	Online: Centimeter Graph Paper Other: Scissors
GROUPING	Work individually or in pairs.

From a sheet of graph paper, cut out squares with areas 9, 16, 25, 36, 49, 64, 81, 100, 121, 144, and 169 square centimeters. Use three squares to construct a triangle as shown.

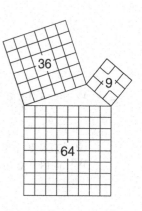

Determine if the triangle is acute, right, or obtuse. If necessary, compare the angles of the triangles to the corner of an index card to determine if the angles are acute, right, or obtuse. Enter the data in Table 1. Use additional sets of three squares to construct other triangles and enter the data in Table 1.

TABLE 1

Area of the Largest Square	Area of the Smallest Square	Area of the Third Square	Sum of the Areas of the Two Smaller Squares	Is the Triangle Acute, Right, or Obtuse?
64	9	36	45	obtuse
100	36	64		
36	16	25		
169				
121				
144				

Use the data from Table 1 to complete Table 2.

TABLE 2

Length of the Longest Side of the Triangle	Square of the Length of the Longest Side	Length of the Shortest Side of the Triangle	Length of the Third Side of the Triangle	Sum of the Squares of the Two Shorter Sides	Is the Triangle Acute, Right, or Obtuse?
8	64	3	6	45	obtuse

Use the data from Table 2 to complete the following statements.

a. If the square of the length of the longest side of a triangle is **less than** the sum of the squares of the lengths of the two shorter sides, the triangle is

a(n) _____ triangle.

b. If the square of the length of the longest side of a triangle is **equal to** the sum of the squares of the lengths of the two shorter sides, the triangle is

a(n) _____ triangle.

c. If the square of the length of the longest side of a triangle is **greater than** the sum of the squares of the lengths of the two shorter sides, the triangle is

a(n) _____ triangle.

Activity 8:   Volume of a Rectangular Solid

PURPOSE	Develop the formula for finding the volume of a rectangular prism.
COMMON CORE SMP	SMP 2, SMP 3, SMP 4, SMP 5, SMP 6, SMP 7
MATERIALS	Other: Set of cubes of the same size
GROUPING	Work individually.
GETTING STARTED	The numbers in the squares in each figure below indicate the number of cubes in a stack. Use your cubes to construct solids made up of these stacks of cubes.

4	3	1
2	1	5
1	2	2

Figure a

8	2
10	4
3	6

Figure b

4	3	6	5
2	5	2	3
4	3	6	4

Figure c

1.   How many cubes are there in **each layer** of each solid and what is the total number of cubes in each solid?

	Layer 1	Layer 2	Layer 3	Layer 4	Layer 5					TOTAL
a.	___	___	___	___	___					_____
b.	___	___	___	___	___	___	___	___	___	_____
c.	___	___	___	___	___					_____

2.   How many more cubes must be added to each solid to construct a rectangular prism? The base and height of the prism must be the same as the base and the maximum height of the solid.

   a. _____   b. _____   c. _____

3.   What is the volume of each rectangular prism in Exercise 2 (the total number of cubes)?

   a. _____   b. _____   c. _____

Use the information in the table to construct rectangular prisms with the given dimensions and complete the table.

Length	Width	Height	Number of Cubes
3	5	2	
4	6	5	
4	4	4	
7	8	3	
6	5.5	10	
14	3	6.5	
15	9	5	
8.7	13	5	

1. Write a formula to find the volume of a rectangular prism using the length, width, and height.

2. Write a rule to find the total **number of cubes** for each prism using the **area of the base** and the **height** of the prism.

3. A rectangular prism with dimensions 2, 5, and 8 can have three different bases, each with different dimensions.

   a. Find the area of each base.

   b. Use each base to find the volume of the rectangular prism.

4. Explain how the formula you wrote in Exercise 2 can be applied to determine the volume of a **cylinder**.

Activity 9: Pyramids and Cones

PURPOSE Develop the relationship between the volume of a pyramid and its related prism and the relationship between a cone and its related cylinder.

COMMON CORE SMP SMP 3, SMP 4, SMP 5, SMP 6, SMP 7

MATERIALS Online: Nets for the Prism, Pyramid, Cylinder, and Cone
Other: Scissors, tape, and rice

GROUPING Work individually or in pairs.

GETTING STARTED Make copies of the nets for the prism, pyramid, cylinder, and cone on construction paper or card stock. Cut out each net and construct each model.

---

1. What is true about the areas of the bases of the prism and the pyramid?

2. What is true about the heights of the prism and the pyramid?

3. Estimate how many pyramids full of rice it will take to fill the prism.

Fill the pyramid with rice. Pour the rice into the prism. Repeat the process until the prism is full. (Be sure to record each time you pour rice into the prism.)

4. One prism full of rice = _____ pyramids full of rice.

5. Write a ratio that compares the volume of the pyramid to the volume of the prism.

6. On the basis of this exploration and the ratio you wrote in Exercise 5, write a rule to determine the volume of a pyramid based on the formula for the volume of a prism.

Repeat the experiment using the cylinder and the cone. Place the cylinder in a box or some container, since it is open at both ends.

1. Write a ratio that compares the volume of the cone to the volume of the cylinder.

2. On the basis of this exploration and the ratio you wrote in Exercise 1, write a rule to determine the volume of a cone based on the formula for the volume of a cylinder.

Activity 10: Surface Area

PURPOSE	Develop the concept of surface area of solids.
COMMON CORE SMP	SMP 3, SMP 4, SMP 5, SMP 6, SMP 7
MATERIALS	Online: Nets for Solids Other: Scissors and tape
GROUPING	Work in pairs.
GETTING STARTED	Make a copy of the nets for the solids on construction paper. Cut out the nets. Then fold them and tape them together to make the solids.

Describe the geometric solid that is formed with each net.

1. Net A

2. Net B

3. Net C

4. Net D

Find the surface area of the solid formed with each net. Explain your method.

1. Net A

2. Net B

3. Net C

4. Net D

EXTENSION	Make three copies of Net B. Cut them out. Then fold them and construct the pyramids. Put the three pyramids together to form a prism. What is the volume of the prism? What is the volume of each pyramid?

Chapter Summary

Perimeter, area, and volume are important concepts in geometry and have wide application in the real world. In fact, of all the topics studied in geometry, they are the source of more problems in everyday life than any other topic.

Research literature contains many references to the need for developmentally appropriate instruction about measurement and the concepts of perimeter, area, and volume. Students not only need to know *how* to measure, they must also know *what it is* that must be measured. Students who learn formulas without a firm grasp of the basic concepts of measurement will have difficulty in their future study of geometry.

Activity 1 introduced area as covering with a given unit. Using a square as the unit of area and determining the number of squares that would cover the different shapes helped develop spatial visualization and estimation skills. You also investigated the relationship between perimeter and area, a concept that is often misunderstood. Many people believe that as the area of a figure increases so does its perimeter. After completing this activity, you discovered that this assumption is false.

Activities 2–4 developed the formulas needed to determine the areas of triangles and special quadrilaterals. Each new polygon was related to one previously studied. Through these activities, you found that all these area formulas have their foundation in the rectangular arrays used to illustrate multiplication.

Activities 5 and 6 introduced one of the most important theorems of geometry, the *Pythagorean theorem*. In Activity 5, you constructed a square on the hypotenuse of a right triangle using the pieces of the squares constructed on one of the other two sides. This illustrated the relationship between the area of the square on the hypotenuse and the sum of the areas of the squares on the other two sides.

Activity 6 provided an opportunity to explore an apparent magic trick. A square was divided into four parts and rearranged. In one case a unit of area disappeared; in the second, an additional unit of area was added. Conservation of area tells us that neither of these situations is possible. Through the application of the Pythagorean theorem you discovered that the eye was deceived when the parts of the square were rearranged.

In Activity 7, you explored the converse of the Pythagorean theorem by putting squares together to form triangles. This also reinforced the classification of triangles by angle measurements and the Triangle Inequality.

Activities 8 and 9 developed the concept of volume and the formulas for the volumes of certain solids. The process of finding a systematic method for counting cubes to determine volume was directly related to multiplying the dimensions of a rectangular array to count the squares when determining area. By comparing a pyramid to a prism and a cone to a cylinder, you developed a connection among these solids and the formulas for finding their volumes. This comparison strategy is the same one used to find area formulas by comparing the areas of related polygons.

In Activity 10, you applied the ideas about area and the formulas for finding the areas of polygons developed in Activities 1–4 to find the surface area of solids.

Throughout this chapter, you were involved in doing mathematics. You discovered patterns, made and tested conjectures, and then reasoned to develop formulas for perimeter, area, and volume. You also constructed new knowledge or reinforced prior knowledge about these important attributes. Your active involvement in developing the formulas helped enhance your understanding of these concepts.

Chapter 11
Transformations, Symmetries, and Tilings

"Transformational geometry offers another lens through which to investigate and interpret geometric objects. To help them form images of shapes through different transformations, students can use physical objects, figures traced on tissue paper, mirrors or other reflective surfaces, figures drawn on graph paper, and dynamic geometry software. They should explore the characteristics of flips, turns, and slides and should investigate relationships among compositions of transformations. These experiences should help students develop a strong understanding of line and rotational symmetry, scaling, and properties of polygons."

—Principles and Standards for School Mathematics

In this chapter, you will study the properties of a class of functions called isometries. An isometry is a mapping or transformation that preserves the distance between points. The transformations in this chapter include slides (translations), flips (reflections), turns (rotations), and glide reflections. You will also investigate the properties of dilations, size transformations that reduce or enlarge figures.

The chapter concludes with an exploration of symmetry—line symmetry, point symmetry, and rotational symmetry. Isometries, dilations, and symmetry are powerful tools for studying congruence, similarity, and other geometric concepts, such as tilings of the plane.

Correlation of Chapter 11 Activities to the
Common Core Standards of Mathematical Practice

Activity Number and Title	Standards of Mathematical Practice
1: Reflections	SMP 2, SMP 3, SMP 4, SMP 5, SMP 6, SMP 7
2: Glide Reflections	SMP 4, SMP 5, SMP 7
3: Translations	SMP 2, SMP 4, SMP 5, SMP 7
4: Rotations	SMP 2, SMP 3, SMP 5, SMP 7
5: Dilations	SMP 3, SMP 4, SMP 5, SMP 7
6: Draw It	SMP 4, SMP 5, SMP 7
7: Tessellations	SMP 2, SMP 3, SMP 4, SMP 5, SMP 6, SMP 7

Activity 1: Reflections

PURPOSE	Explore reflections and their properties.
COMMON CORE SMP	SMP 2, SMP 3, SMP 4, SMP 5, SMP 6, SMP 7
MATERIALS	Other: A centimeter ruler, a protractor, and a Mira™
GROUPING	Work individually or in pairs.
GETTING STARTED	A Mira is a plastic drawing device that acts like a mirror. A Mira reflects objects, but since it is transparent, the image of an object reflected in it also appears behind the Mira.

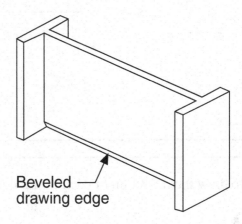

Beveled drawing edge

The drawing edge of a Mira is beveled. When using a Mira, place it with the beveled edge down. Look directly through the Mira from the side with the beveled edge to locate the image of the object behind the Mira.

Place your Mira so that the image of circle *A* fits on circle *B*. Hold the Mira steady with one hand and draw a line along the drawing edge.

Take away the Mira. The line you have drawn is the *line of reflection*. It represents the Mira. How does the line of reflection appear to be related to points *A* and *B*?

For each pair of figures below, use a Mira to fit the image of one of the figures onto the other. Then draw the line of reflection.

1.
2.
3.

Use a Mira to draw the reflection of each figure through the given line.

1.
2.

3.
4.

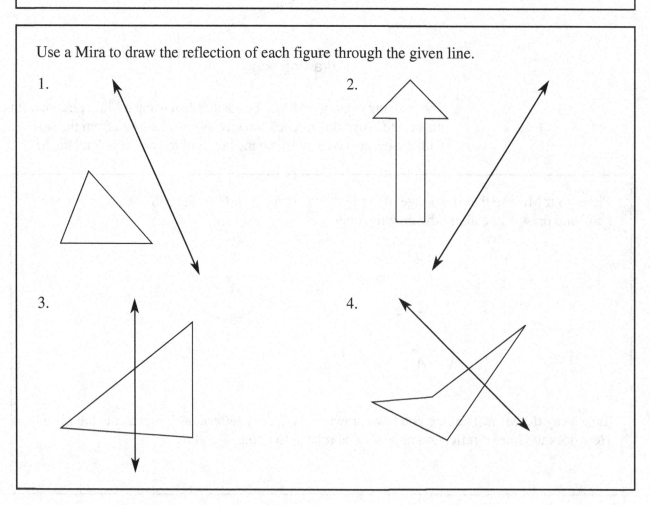

Use a Mira to mark the location of the reflection of each point through the line ℓ. Use prime notation to name each image point. For example, the image of point D would be named D'.

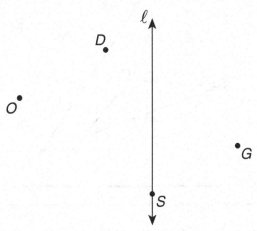

Draw line segments $\overline{DD'}$, $\overline{OO'}$, and $\overline{GG'}$. Label the points where line ℓ intersects these segments C, A, and T, respectively.

1. What is the relationship between the line ℓ and the segments $\overline{DD'}$, $\overline{OO'}$, and $\overline{GG'}$?

2. Where is point S located in relation to line ℓ?

3. What is the relationship between the point S and its reflection S'?

Use a protractor and ruler to make the following constructions.

1. A line ℓ so that point P is the reflection of point A through ℓ.

2. A line ℓ so that pentagon R is the reflection of pentagon S through ℓ.

P •

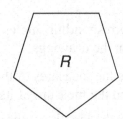

• A

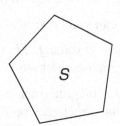

3. a. Explain how you constructed line ℓ.

    b. Why did you construct line ℓ as you did?

Use a ruler and protractor to construct the reflection of $\triangle ABC$ through line ℓ.

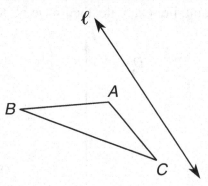

Use a Mira to reflect the figure *FLAG* through the line ℓ. Use prime notation to label the reflection.

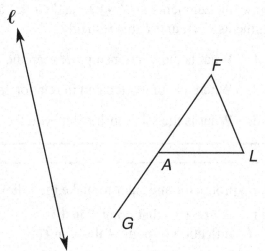

Measure the corresponding angles and segments in *FLAG* and its reflection image. Indicate the measures on the drawings.

1. Examine the measures of the angles. What can you conclude about the measure of an angle and the measure of its reflection?

2. Examine the lengths of the segments. What can you conclude about the length of a segment and the length of its reflection?

3. What can you conclude about a triangle and its reflection through a line?

4. a. Imagine tracing $\triangle FAL$ from *F* to *A* to *L* and back to *F*. What direction (clockwise or counterclockwise) would you move?

   b. Now imagine tracing the image $\triangle F'A'L'$ from *F'* to *A'* to *L'* and back to *F'*. What direction would you move?

   c. How does reflecting a figure through a line affect its orientation?

Activity 2: Glide Reflections

PURPOSE	Explore glide reflections and their properties.
COMMON CORE SMP	SMP 4, SMP 5, SMP 7
MATERIALS	Other: A compass, a centimeter ruler, a protractor, and a Mira™
GROUPING	Work individually or in pairs.
GETTING STARTED	In the figure below, footprint F_3 is a glide reflection of footprint F_1. As the name suggests, a glide reflection is a glide (or translation) followed by a reflection. However, the reflecting line must be parallel to the direction of the glide.

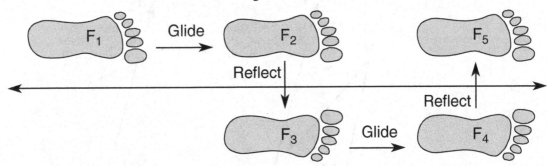

1. A glide reflection maps $\triangle ABC$ onto $\triangle A'B'C'$. Draw segments AA', BB', and CC' and find their midpoints. What appears to be true about the midpoints of the segments?

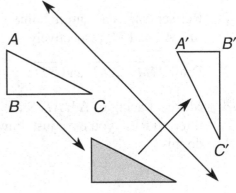

2. A glide reflection maps $\triangle XYZ$ onto $\triangle X'Y'Z'$. Find the reflecting line and draw the glide image of $\triangle XYZ$.

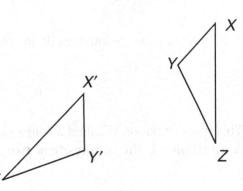

3. Study the figure at the top of the page. What single transformation is equivalent to the result of using a glide reflection twice?

Activity 3: Translations

PURPOSE	Explore translations and their properties.
COMMON CORE SMP	SMP 2, SMP 4, SMP 5, SMP 7
MATERIALS	Other: A centimeter ruler, a piece of tracing paper, and a Mira™
GROUPING	Work individually or in pairs.

1. Reflect $\triangle MAT$ through line ℓ_1. Use prime notation to label the reflection.

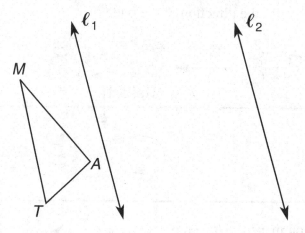

2. Reflect $\triangle M'A'T'$ through line ℓ_2. Let M'', A'', and T'' denote the images of M', A', and T', respectively.

3. Draw $\overline{MM''}$, $\overline{AA''}$, and $\overline{TT''}$.

4. Make a tracing of $\triangle MAT$. Slide it onto $\triangle M''A''T''$ by moving its vertices along the three "tracks" you have just drawn. Is it necessary to flip or to turn the tracing to do this?

5. What two relationships do the tracks appear to have?

This transformation is called a *translation*. The exercises illustrate the following definition: **A *translation* is the composite of two reflections through parallel lines.**

1. Translate $\overline{ID}$ by reflecting it through line ℓ_3 and then reflecting the image through line ℓ_4. Let $\overline{ES}$ denote the final translation image of $\overline{ID}$ (E is the image of I and S is the image of D).

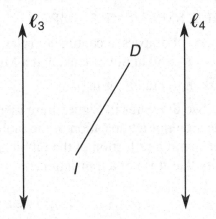

2. What is the distance between lines ℓ_3 and ℓ_4?

3. Measure IE and DS. How do these lengths compare with your answer in Exercise 2?

4. In what direction was $\overline{ID}$ translated?

5. a. Now translate $\overline{ID}$ by reflecting it through line ℓ_4 and then reflecting the image through line ℓ_3. Let $\overline{GL}$ denote the translation image of $\overline{ID}$ (G is the image of I and L is the image of D).

   b. Measure the lengths IG and DL. How do these lengths compare with the result in Exercise 2?

   c. In what direction was $\overline{ID}$ translated?

6. a. If a point is translated by reflecting it through two parallel lines that are x units apart, what is the distance between the point and its image?

   b. If a figure is translated by first reflecting it through line ℓ_5 and then reflecting its image through line ℓ_6, in what direction will the figure be translated?

   c. Use the translation of $\triangle MAT$ on the previous page to check your conclusions in Parts a and b.

Activity 4:   Rotations

PURPOSE	Explore rotations and their properties.
COMMON CORE SMP	SMP 2, SMP 3, SMP 5, SMP 7
MATERIALS	Other: A compass, a centimeter ruler, a protractor, tracing paper, a straight pin or tack, and a Mira™
GROUPING	Work individually or in pairs.
GETTING STARTED	The two drawings in Figure 1 are identical, but the duck is neither a reflection nor a translation of the rabbit. Use a Mira to verify that the duck is not a reflection of the rabbit and a tracing of the duck to verify that it is not a translation.

Place a piece of tracing paper over the rabbit in Figure 1. Pin it at point *P*. Trace the rabbit and then turn the paper about the pin until the rabbit coincides with the duck. This illustrates why the duck is called a *rotation image* of the rabbit.

P.

Figure 1

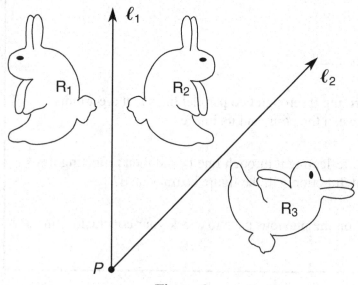

P

Figure 2

The drawings in Figure 2 show that if the rabbit, R_1, is reflected through line ℓ_1 and its image, R_2, is reflected through line ℓ_2, then the result is the duck, R_3. Verify this with your Mira.

This example shows that the rotation that transforms the rabbit into the duck is the composite of two reflections through intersecting lines. This illustrates the following definitions:

A *rotation* is the composite of two reflections through intersecting lines.

The *center of rotation* is the point at which the two lines intersect.

1. What is the measure of the acute angle formed by lines ℓ_1 and ℓ_2?

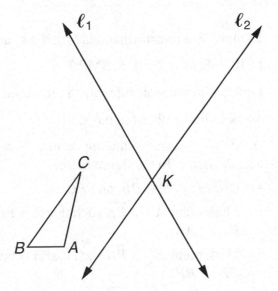

2. Rotate $\triangle ABC$ by reflecting it through line ℓ_1 and then reflecting its image through line ℓ_2. Let X, Y, and Z be the rotation images of points A, B, and C, respectively.

   a. Draw and measure $\angle BKY$. Note that the vertex of $\angle BKY$ is at the center of rotation and that one side of the angle contains a point B on the original figure. The other side contains the rotation image Y of point B. The measure of an angle formed this way is the *magnitude of the rotation*.

   b. How does the magnitude of the rotation compare with your answer to Exercise 1?

   c. In what direction, clockwise or counterclockwise, was $\triangle ABC$ rotated?

3. Rotate $\triangle ABC$ by reflecting it through line ℓ_2 and then reflecting its image through line ℓ_1. Let R, S, and T be the rotation images of points A, B, and C, respectively.

   a. What is the magnitude of the rotation? How does it compare with your answer to Exercise 1?

   b. In what direction was $\triangle ABC$ rotated?

4. a. On the figure in Exercise 1, draw three circles with center K and radii $\overline{BK}$, $\overline{CK}$, and $\overline{AK}$, respectively. What do you notice about the circles?

   b. Make a tracing of $\triangle ABC$. Slide it onto $\triangle XYZ$ by moving its vertices along the circular "tracks" you have just drawn. Is it necessary to flip or to turn the tracing?

5. a. A point is rotated by reflecting it through two intersecting lines that form an acute angle with measure $x°$. What will be the magnitude of the rotation?

   b. If a figure is rotated by first reflecting it through line ℓ_3 and then reflecting its image through line ℓ_4, in what direction will the figure be rotated?

Activity 5:   Dilations

PURPOSE	Explore size transformations (dilations) and their properties.
COMMON CORE SMP	SMP 3, SMP 4, SMP 5, SMP 7
MATERIALS	Other: A centimeter ruler and a protractor
GROUPING	Work individually or in pairs.
GETTING STARTED	A *dilation* is a tranformation defined by a point called the *center* and a *scale factor*. In the figure below:

- Draw rays $\overrightarrow{PA}$, $\overrightarrow{PB}$, and $\overrightarrow{PC}$.

- Mark point A' on $\overrightarrow{PA}$ such that A is between P and A' and $PA = AA'$.

- Mark point B' on $\overrightarrow{PB}$ such that B is between P and B' and $PB = BB'$.

- Mark point C' on $\overrightarrow{PC}$ such that C is between P and C' and $PC = CC'$.

- Draw $\triangle A'B'C'$. $\triangle A'B'C'$ is the *image* of $\triangle ABC$.

Q.

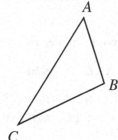

P.

1. Measure the sides and angles of $\triangle ABC$ and $\triangle A'B'C'$. Record the measures in the tables.

△ABC	
m∠A	
m∠B	
m∠C	
AB	
AC	
BC	

△A′B′C′	
m∠A′	
m∠B′	
m∠C′	
A′B′	
A′C′	
B′C′	

2. a. Are $\triangle ABC$ and $\triangle A'B'C'$ the same size? Explain.

    b. Do the triangles have the same shape? Justify your answer.

3. Find the following ratios. What do you notice?

    $$\frac{A'B'}{AB} =$$

    $$\frac{A'C'}{AC} =$$

    $$\frac{B'C'}{BC} =$$

4. a. What is the ratio of PA' to PA?

    b. How does this ratio compare to the ones in Exercise 3?

    c. Point P is the center of the dilation and the ratio $PA' : PA$ is the scale factor. Why does it make sense to call this ratio a scale factor?

5. We say two figures are *similar* if they have the same shape but not necessarily the same size. Use your results from Exercises 1–4 to write a definition of similar triangles.

6. a. Find the midpoints of segments $\overline{A'B'}$, $\overline{B'C'}$, and $\overline{C'A'}$, and label them R, S, and T, respectively. Draw $\triangle RST$.

    b. Drawing $\triangle RST$ divided $\triangle A'B'C'$ into four triangles. How do these triangles appear to be related?

    c. How is $\triangle RST$ related to $\triangle ABC$?

    d. How is the area of $\triangle A'B'C'$ related to the area of $\triangle ABC$?

7. In the figure at the beginning of the activity:

    -   Mark point A'' on $\overrightarrow{PA}$ such that A is between P and A'' and $AA'' = 2 \cdot PA$.

    -   Mark point B'' on $\overrightarrow{PB}$ such that B is between P and B'' and $BB'' = 2 \cdot PB$.

    -   Mark point C'' on $\overrightarrow{PC}$ such that C is between P and C'' and $CC'' = 2 \cdot PC$.

    -   Draw $\triangle A''B''C''$.

8. a. Find the ratio of the length of each side of $\triangle A''B''C''$ to the length of the corresponding side in $\triangle ABC$.

    b. Is $\triangle ABC$ similar to $\triangle A''B''C''$? Explain.

    c. What is the scale factor of the dilation (the ratio of PA'' to PA)? How does it compare to the ratios in Part a?

    d. Trisect the sides of $\triangle A''B''C''$. Show how you could connect the trisection points to show that the area of $\triangle A''B''C''$ is nine times the area of $\triangle ABC$.

9. a. Is $\triangle A''B''C''$ similar to $\triangle A'B'C'$?

    b. What is the scale factor of the dilation that maps $\triangle A'B'C'$ to $\triangle A''B''C''$ (the ratio of PA'' to PA')?

    c. How is the area of $\triangle A''B''C''$ related to the area of $\triangle A'B'C'$?

10. In the figure at the beginning of the activity:

    • Draw rays $\overrightarrow{QA}$, $\overrightarrow{QB}$, and $\overrightarrow{QC}$.

    • Mark point X on $\overrightarrow{QA}$ such that A is between Q and X and $QA = AX$.

    • Mark point Y on $\overrightarrow{QB}$ such that B is between Q and Y and $QB = BY$.

    • Mark point Z on $\overrightarrow{QC}$ such that C is between Q and Z and $QC = CZ$.

    • Draw $\triangle XYZ$.

11. What effect does moving the center of dilation have on the result of the transformation?

12. A dilation maps $\triangle JKL$ to $\triangle J'K'L'$. If the center is M:

    a. What is the scale factor of the dilation?

    b. How are the lengths of the sides of the two triangles related?

    c. How are the areas of the two triangles related?

Activity 6: Draw It

PURPOSE	Apply the concept of symmetry to drawing figures.
COMMON CORE SMP	SMP 4, SMP 5, SMP 7
MATERIALS	Online: Centimeter Graph Paper Other: Either a mirror or a Mira™
GROUPING	Work individually.

1. Use the grid to complete the drawing of the figure at the right. The final figure should be symmetric with respect to the vertical line.

2. Place a mirror or a Mira on the line so that you can see the reflection of the left half of the picture in it. What do you observe?

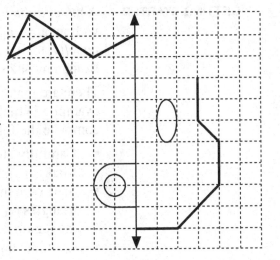

1. Use the grid to complete the drawing of the figure at the right. The final figure should be symmetric with respect to the point *O*.

2. Rotate the page 180° and look at the figure. What do you observe?

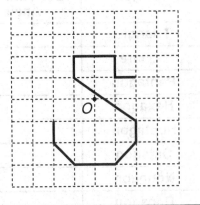

EXTENSIONS

1. Draw a figure that has line symmetry but not rotational symmetry.

2. Draw a figure that has point symmetry but not line symmetry.

3. Draw a figure that has rotational symmetry but neither point nor line symmetry.

4. Make up a design that has one or more lines of symmetry. Draw part of it on graph paper and have a classmate complete the design using symmetries as in the activity above.

Activity 7:   Tessellations

PURPOSE	Construct tilings using regular polygons and explore regular and semi-regular tessellations.
COMMON CORE SMP	SMP 2, SMP 3, SMP 4, SMP 5, SMP 6, SMP 7
MATERIALS	Online: Regular Polygons (Printing the page on card stock works well.)
	Other: Scissors, tape, a ruler, and five index cards
GROUPING	Work individually or in groups of 2 or 3.
GETTING STARTED	A *tessellation* is a tiling that uses congruent figures to cover the plane without any gaps or overlaps. The term tessellation comes from *tessella*, the Latin word for the small square tiles used in ancient Roman mosaics.

There are many interesting questions related to tessellations. For example: **Which regular polygons will tessellate?** To find out, complete the following experiment.

- Carefully cut out the regular polygons and sort them by shape.
- Mark a point *P* near the center of a separate sheet of paper.
- Start with the equilateral triangles. Place a vertex of one of the triangles on *P*. Continue placing triangles edge-to-edge to cover the region around *P*. Record your results in the table.
- Continue this process until you have tested each regular polygon.

Polygon	Measure of Each Interior Angle	Number of Angles at *P*	Sum of the Measures of the Angles at *P*	Does the Polygon Tessellate?
Triangle				
Square				
Pentagon				
Hexagon				
Heptagon				
Octagon				
Nonagon				
Decagon				
Dodecagon				

1. What must be true about the measure of an interior angle of a regular polygon in order for the polygon to tessellate?

2. a. Which of the regular polygons tessellated?

    b. Make a sketch to show that each polygon in Part a will tile the plane. These are the *regular* tessellations.

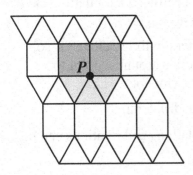

3. Use your result from Exercise 1 to show that the polygons in Exercise 2 are the only regular polygons that will tessellate.

In the experiment, you discovered how one regular polygon can be used to construct a tessellation. Tessellations can also be made using combinations of regular polygons. For example, the tessellation at the left was made using two squares and three equilateral triangles.

4. Find a different way to arrange two squares and three equilateral triangles around a point *P* to create a tessellation. Make a sketch of the tessellation.

5. Construct four more tessellations that include more than one type of regular polygon. Make a sketch of each one.

A tessellation that uses a combination of regular polygons joined edge-to-edge and has the same arrangement of polygons around every vertex is *semi-regular*. The tessellation above is semi-regular, but the one below is not since the arrangements of polygons at vertices *Q* and *R* are different.

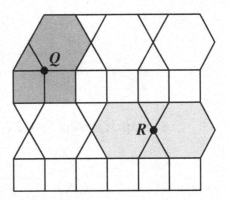

6. Which of the tessellations you created in Exercises 4 and 5 are semi-regular? If a tessellation isn't semi-regular, explain why not.

Now let's explore tessellating the plane with irregular polygons.

7. a. Use a ruler to draw a non-equilateral triangle on an index card and cut it out.

   b. Trace around your triangle to create a tessellation. Be sure corresponding sides of the triangles are joined edge-to-edge.

   c. Compare your triangle and tessellation with those of several classmates. Were there any triangles that would not tessellate?

   d. Will any triangle tessellate? If so, show how to arrange the triangles around a point to create a tessellation.

8. a. Draw a convex quadrilateral on an index card and cut it out.

   b. Trace around the quadrilateral to construct a tessellation. Be sure to join the quadrilaterals edge-to-edge.

   c. Do you think any convex quadrilateral will tessellate? If so, show how to arrange the quadrilaterals around a point to create a tessellation. If not, explain why not.

9. Repeat Exercise 8 using a concave quadrilateral.

EXTENSIONS Follow the steps below to learn how translations and rotations can be used to construct tessellations.

Using Translations:

Step 1: Draw a parallelogram on an index card and cut it out.

Step 2: Draw a shape along one side of the parallelogram. Cut it out and translate it to the opposite side. Tape the translated piece on the opposite side as shown.

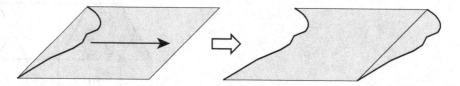

Step 3: Repeat Step 2 on the other two parallel sides.

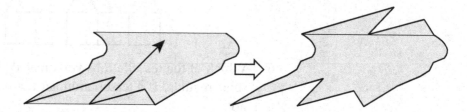

Step 4: Use your shape to construct a tessellation.

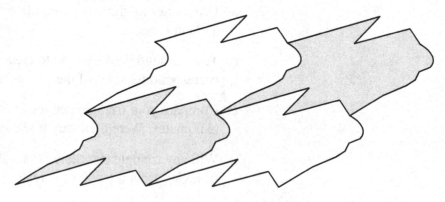

Using Rotations:

Step 1: Draw an equilateral triangle on an index card and cut it out.

Step 2: Draw a shape along one side of the triangle. Cut it out and rotate it 60° about an endpoint to modify a second side. Tape on the rotated piece.

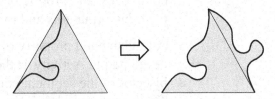

Step 3: Find the midpoint of the remaining side. Draw a shape along one half of the side. Cut out the shape, rotate it 180° about the midpoint, and tape it to the other half of the side.

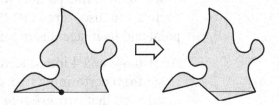

Step 4: Use your shape to construct a tessellation.

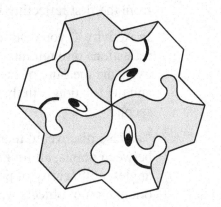

Chapter Summary

In Activities 1–4, you studied four isometries—translations, reflections, rotations, and glide reflections—and discovered some of their properties. A variety of tools—tracing paper, Miras, rulers, compasses, and protractors—were used in the activities to perform the transformations and to study their properties.

A fundamental property of *reflections*, or mirror images, was developed in Activity 1: the reflecting line is the perpendicular bisector of the segment connecting a point and its reflection image. This property makes it possible to reflect objects using ruler and compass constructions or coordinate methods.

In Activity 2, you explored glide reflections. A *glide reflection* is the composite of two transformations, a glide (translation) followed by a reflection in a line parallel to the direction of the glide, the glide vector. You discovered that the midpoint of the segment joining a point and its image under a glide reflection lies on the reflecting line.

Activities 3 and 4 introduced the idea that translations and rotations result from performing two successive reflections. In Activity 3, you discovered that a *translation* is the composite of two reflections through parallel lines. The distance the object is translated is twice the distance between the lines and the translation is in the direction from the first reflecting line to the second.

In Activity 4, you discovered that a *rotation* is the composite of two reflections through intersecting lines. The magnitude of the rotation is twice the measure of the acute angle formed by the reflecting lines and the rotation is in the direction from the first reflecting line to the second.

You also discovered that in addition to preserving the distance between points, all four transformations preserved the measure of angles, collinearity of points, and congruence of objects. The orientation of objects was preserved by translations and rotations, but reversed by reflections and glide reflections.

In Activity 5, you explored size transformations, or *dilations*, and saw how they enlarge or reduce a figure. A dilation is defined by a point called the center, and a scale factor. Dilations do not preserve the distance between points; however, they do preserve collinearity of points and the measures of angles. A figure and its dilation image are similar and, if the scale factor is k, the area of the image is k^2 times the area of the original figure.

Activity 6 explored the concept of symmetry. There are three kinds of symmetry. If an object has *line symmetry*, it can be divided into two halves in such a way that each half is the mirror image of the other.

An object has *rotational symmetry* if it can be made to coincide with itself by a rotation of less than 360° about a point. Objects that have a rotational symmetry of 180° are said to have *point symmetry*. If an object has point symmetry, it appears the same when viewed right-side-up or up-side-down.

Line and rotational symmetry are very common in nature. The body structure of most animals, including humans, possesses line symmetry. Some trees and the leaves of many plants also exhibit line symmetry, while many flowers exhibit rotational symmetries.

You used geometric transformations in Activity 7 to study tessellations, tilings that use congruent shapes to completely cover the plane without any gaps or overlaps. You discovered that only three regular polygons—equilateral triangles, squares, and hexagons—will tessellate and that **all** triangles and quadrilaterals tessellate.

Chapter 12
Congruence, Constructions, and Similarity

"... students also need experience in working with congruent and similar shapes ... they should understand that congruent shapes and angles are identical and can be "matched" by placing one atop the other. Students can begin with an intuitive notion of similarity: similar shapes have congruent angles but not necessarily congruent sides ... they should extend their understanding of similarity to be more precise, noting for instance, that similar shapes "match exactly when magnified or shrunk" or that their corresponding angles are congruent and their corresponding sides are related by a scale factor."
—*Principles and Standards for School Mathematics*

"They also solve problems about similar objects (including figures) by using scale factors that relate corresponding lengths of the objects or by using the fact that relationships of lengths within an object are preserved in similar objects. ... Students apply this reasoning about similar triangles to solve a variety of problems including those that ask them to find heights and distances."
—*Curriculum Focal Points for Prekindergarten through Grade 8 Mathematics*

The activities in this chapter will engage you in problem-solving situations in which you have the opportunity to explore and discover geometric concepts and relationships. You will construct triangles given a combination of parts and compare the results with that of a classmate. Then, you will be asked to make and test conjectures based on your observations.

You will explore the concept of similarity using pattern blocks to construct and compare similar polygons. Then, you will use a protractor and ruler to construct triangles given certain sets of measures for the angles and check to determine if the triangles are similar. In one activity, you will use a computer drawing program to complete a construction and then alter the figure dynamically to illustrate that a conclusion is true for any triangle.

In another activity, you will use the concepts that were developed in the previous activities on similar triangles to measure inaccessible heights. This activity demonstrates the use of similar triangles in a real-world application. It provides a good answer to the age-old question, "when are we ever going to use this?"

Correlation of Chapter 12 Activities to the Common Core Standards of Mathematical Practice

Activity Number and Title	Standards of Mathematical Practice
1: Triangle Properties—Sides	SMP 2, SMP 3, SMP 4, SMP 5, SMP 6, SMP 7
2: To Be or Not to Be Congruent?	SMP 2, SMP 3, SMP 4, SMP 5, SMP 6, SMP 7
3: Pattern Block Similarity	SMP 2, SMP 3, SMP 4, SMP 5, SMP 6, SMP 7
4: Similar Triangles	SMP 3, SMP 4, SMP 5, SMP 7
5: Outdoor Geometry	SMP 2, SMP 3, SMP 4, SMP 5, SMP 6, SMP 7, SMP 8
6: Side Splitter Theorem	SMP 3, SMP 4, SMP 5, SMP 7, SMP 8

Activity 1: Triangle Properties—Sides

PURPOSE	Develop the Triangle Inequality, reinforce construction of a triangle using a compass and ruler, and introduce congruence of triangles.
COMMON CORE SMP	SMP 2, SMP 3, SMP 4, SMP 5, SMP 6, SMP 7
MATERIALS	Other: A centimeter ruler, a compass, a geoboard, and geo-bands
GROUPING	Work in pairs.
GETTING STARTED	Make all constructions on a separate piece of paper. As you construct the triangles in each section, compare your results with those of your partner.

1. Use a ruler to construct two different triangles in which one side is 6 cm long and another side is 9 cm long.

    a. Compare your triangles with your partner's triangles. What do you notice?

    b. How many different triangles could you construct given the lengths of two sides?

2. Use a ruler and a compass to construct triangles with sides of the following lengths.

    a. 9 cm, 7 cm, 5 cm   b. 7 cm, 7 cm, 10 cm   c. 8 cm, 8 cm, 8 cm

    d. 6 cm, 5 cm, 12 cm   e. 7 cm, 12 cm, 11 cm   f. 10 cm, 6 cm, 4 cm

3. Which sets of measures **did not** result in a triangle? Why is it impossible to construct triangles with sides of these lengths?

4. Which sets of measures **did** result in a triangle? In these cases, what is true about the sum of the lengths of the two shorter sides?

5. List measures for sets of three side lengths that **can** be used to construct a triangle. Do not use a set in which all the lengths are equal.

    a. _____   b. _____   c. _____

6. List measures for sets of three side lengths that **cannot** be used to construct a triangle.

    a. _____   b. _____   c. _____

7. What can you conclude about the lengths of the sides of a triangle?

1. For each set of three measures in Exercise 2 that resulted in a triangle:

   a. Classify the triangle using a combination of classifications, one by sides and one by angles.

   b. Compare each of your triangles to the corresponding triangle of your partner. What do you notice?

2. With your partner, decide on two additional sets of measures for three segments that **will** form a triangle. Each person should construct a triangle using these numbers. Compare the triangles as before. What do you notice?

How many unique triangles can be constructed on a 3 peg × 3 peg square portion of a geoboard? Build your triangles on a geoboard and record the results on the grids below. Classify each triangle that you construct. It may be necessary to put more than one triangle on a grid.

Activity 2: To Be or Not to Be Congruent?

PURPOSE	Develop the triangle congruence properties using constructions.
COMMON CORE SMP	SMP 2, SMP 3, SMP 4, SMP 5, SMP 6, SMP 7
MATERIALS	Other: Ruler, protractor, and compass
GROUPING	Work in pairs.
GETTING STARTED	In Activity 1, you learned that two triangles are congruent if the three sides of one triangle are congruent respectively to the three sides of the other. Are there other combinations of corresponding parts of two triangles that will ensure that the triangles are congruent?

1. Use a compass and ruler to construct a triangle with the parts given in each exercise.

    a.

    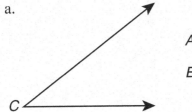

    b.

    c.

    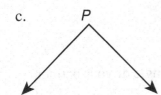
    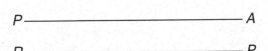

2. In each part of Exercise 1, how were the given sides and the angle of the triangle related?

3. Compare each triangle you constructed to the corresponding triangle of your partner. What do you notice?

4. What can you conclude from your answers to the above questions?

1. In the following problems, use a ruler and compass to construct a triangle with the given parts. Label the third angle of each triangle Z.

    a.

    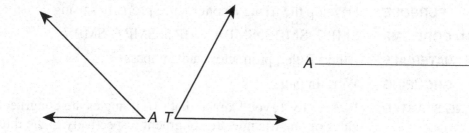

    b.

    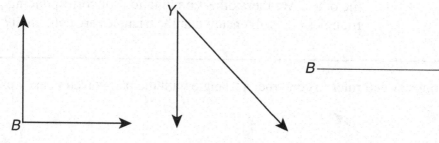

    c.

    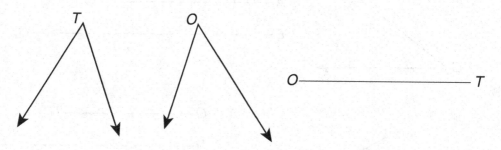

2. In each part of Exercise 1, how is the side related to the given angles?

3. Compare each of your triangles to the corresponding triangle of your partner. What do you notice?

4. What can you conclude from your answers to the above questions?

1. Use a ruler and a protractor to construct a triangle *BAD* in which $AB = 10$ cm, $m\angle A = 40°$, and $m\angle D = 80°$.

   a. How are the angles and the given side of the triangle related?

   b. Describe step by step how you constructed the triangle.

   c. Compare the triangle you constructed with that of your partner. What do you notice?

   d. What can you conclude from Parts a and c?

2. For each of the following, use a ruler, a protractor, and a compass to construct a triangle with the given parts.

   a. $TO = 8$ cm   b. $HA = 10$ cm   c. $TI = 9$ cm

       $OP = 5$ cm $AT = 6$ cm $IE = 4$ cm

       $m\angle T = 30°$ $m\angle H = 37°$ $m\angle T = 40°$

3. For each part in Exercise 2, how are the given sides and angle related?

4. Which exercises did not result in a triangle?

5. Did any of the exercises result in more than one triangle? If so, which one(s)?

6. If any exercise resulted in only one triangle, what type of triangle was it?

7. What can you conclude from your answers in Exercises 3–6?

Activity 3: Pattern Block Similarity

PURPOSE	Develop the concept of similarity using pattern blocks.
COMMON CORE SMP	SMP 2, SMP 3, SMP 4, SMP 5, SMP 6, SMP 7
MATERIALS	Pouch: Pattern Blocks
GROUPING	Work in pairs or individually.
GETTING STARTED	Using squares, you can construct larger squares that are **similar** to the original one.

Example:

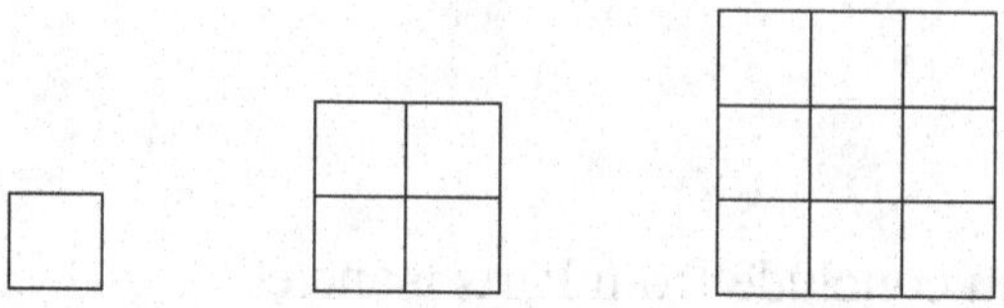

Figure 1
(original block)   **Figure 2**   **Figure 3**

1. Use triangles to construct the next larger triangle similar to the green triangle. Sketch the similar triangle below.

2. For each pattern block, is it possible to use only blocks of the same shape to construct a larger shape similar to the original block? If so, use the least number of blocks possible to construct the similar shape and sketch it below.

3. For each shape in Exercise 2, use the least number of blocks possible to construct the **next larger** similar shape. Sketch the figures below.

4. How many blocks were needed for each figure?

1. For each shape in Exercise 3 on the previous page, record the number of blocks that were used to construct the similar figure in the table below.

Figure	1 (original block)	2	3	4	5	6	7	...	n
Number of Blocks	1	4							...	

2. For each shape in Exercise 3, use the least number of blocks possible to construct the **next larger** similar shape. Record the number of blocks that are used in the table.

3. Describe the set of numbers in the **Number of Blocks** row of the table.

4. Describe the *numerical pattern of the differences* in the number of blocks in each successive similar figure.

5. Use your results in Exercises 3 and 4 to complete the table.

Construct two different parallelograms using four red trapezoids each.

1. Are the measures of the angles of the red parallelograms equal to the measures of the corresponding angles of the blue rhombus?

2. Are the lengths of the corresponding sides proportional?

3. Are the red parallelograms similar to the blue rhombus? Explain.

Construct a trapezoid using three red trapezoids.

1. Are the measures of the angles of the small trapezoid equal to the measures of the corresponding angles of the large trapezoid?

2. Are the lengths of the corresponding sides proportional?

3. Are the trapezoids similar? Explain.

Construct a rhombus using four tan rhombi. Compare this figure to the blue rhombus.

1. Are the lengths of the corresponding sides of the two rhombi proportional? If so, what is the ratio of the corresponding sides?

2. Are the measures of the corresponding angles equal?

3. Are the two rhombi similar? Explain.

Activity 4: Similar Triangles

PURPOSE	Introduce the concept of similarity of triangles and reinforce the construction of a triangle using a protractor and ruler.
COMMON CORE SMP	SMP 3, SMP 4, SMP 5, SMP 7
MATERIALS	Other: A centimeter ruler and a protractor
GROUPING	Work in pairs.
GETTING STARTED	Make all constructions on a separate piece of paper.

1. Use a ruler and a protractor to construct a triangle that has three angles with the indicated measures.

    a. 35°, 55°, 90°   b. 40°, 75°, 65°   c. 130°, 25°, 25°

    d. 60°, 60°, 60°   e. 110°, 50°, 20°   f. 70°, 70°, 40°

2. Where is the longest (shortest) side of the triangle in relation to the largest (smallest) angle in the triangle?

3. Compare each of your triangles with the corresponding triangles of your partner. What do you notice? Explain your answer.

Measure the lengths of the sides of each of your triangles. Then find the ratio of the length of each side of your triangle to the length of the corresponding side of your partner's triangle and record the ratios in the table below.

Problem	Ratios of Lengths of Sides					
	Shortest Side to Shortest Side		Middle Side to Middle Side		Longest Side to Longest Side	
	Fraction	Decimal	Fraction	Decimal	Fraction	Decimal
a						
b						
c						
d						
e						
f						

4. For each pair of triangles, what can you conclude about the ratios of the corresponding sides?

Activity 5: Outdoor Geometry

PURPOSE	Apply the properties of similar triangles to indirect measurement.
COMMON CORE SMP	SMP 2, SMP 3, SMP 4, SMP 5, SMP 6, SMP 7, SMP 8
MATERIALS	Other: A mirror, a 5-meter to 10-meter measuring tape, a straw, a small washer, and a piece of thread 40 cm long
GROUPING	Work in groups of four.
GETTING STARTED	Identify several tall objects to be measured. One student should make a sketch of the method of solution (see example) and record the data. Another student can provide the shadow or do the sighting, as in the mirror method. The other two students can do the measuring. Divide the tasks among the members of the group and switch roles as each successive object is measured.

SHADOW METHOD

Measure *NA*, the height of a person; *AD*, the length of the person's shadow; and *JI*, the length of the shadow of the object for which the height is being determined.

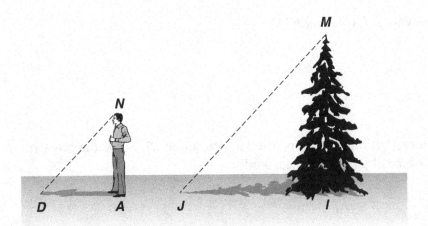

1. Explain why $\triangle DAN \sim \triangle JIM$.

2. Write a proportion that can be used to determine the height of the object, *MI*. Solve the proportion to find *MI*.

MIRROR METHOD

With a felt marker, draw a segment connecting the midpoints of one pair of opposite sides of a mirror. Place the mirror on the ground so that the segment on the mirror is parallel to a line determined by the tips of the toes of the shoes of a person facing the object to be measured. The person should look into the mirror and align the reflection of the top of the object to be measured (point J on the hoop) with the line on the mirror represented by $\overline{RS}$. Point M is the intersection of $\overline{AI}$ and $\overline{RS}$.

Measure AM, MI, and EI. Note that EI is the eye-to-ground distance, not the height of the person. The point I should be vertically below the person's eye (point E), approximately at the toes of the shoes.

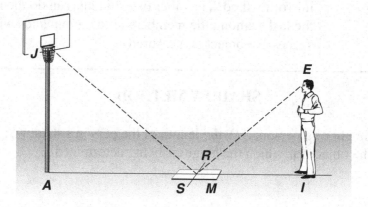

1.   Explain why $\triangle JAM \sim \triangle EIM$.

2.   Write a proportion that can be used to determine JA. Solve the proportion and find the height of the object being measured.

HYPSOMETER METHOD

Use a clipboard with a pad of paper attached to it. Pin a drinking straw along the top of the pad at *C* and *B*. Attach a small weight to one end of a 40-centimeter length of thread and tie the other end of the thread to the pin at *B*. One person should hold the clipboard and sight through the straw until the top of the object to be measured is sighted. A second person should mark the point *F* on the edge of the pad to determine $\triangle BAF$. Measure *DC*, *BA*, *AF*, and the distance from eye level to the ground.

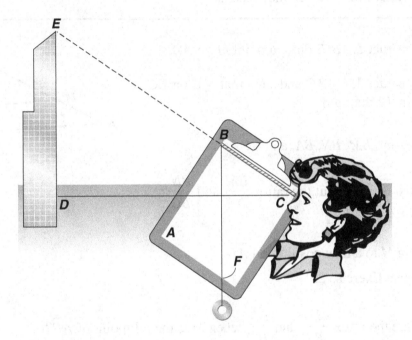

1. Explain why $\triangle CDE \sim \triangle BAF$.

2. Write a proportion that could be used to determine *DE*. Solve the proportion and find *DE*.

3. Is *DE* equal to the measure of the height of the object being measured? What must be done to *DE* to determine the height of the object?

Activity 6: Side Splitter Theorem

PURPOSE	Use a computer and geometry drawing software to explore similar triangles and the Side Splitter Theorem.
COMMON CORE SMP	SMP 3, SMP 4, SMP 5, SMP 7, SMP 8
MATERIALS	Other: Computer and geometry drawing software
GROUPING	Work individually.

1. a. Construct $\triangle ABC$. Place a point M on $\overline{AB}$.

    b. Construct $\overline{MN} \,\|\, \overline{AC}$ and such that N is on $\overline{BC}$ as in the diagram.

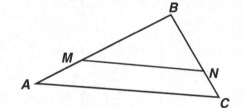

2. a. Measure BM, BN, BA, and BC.

    b. What is true about the ratios $\dfrac{BM}{BA}$ and $\dfrac{BN}{BC}$?

3. a. Drag M to a new position on $\overline{AB}$.

    b. Repeat Exercise 2.

4. What are the ratios $\dfrac{BM}{BA}$ and $\dfrac{BN}{BC}$ when M is the midpoint of $\overline{AB}$?

5. a. Click on B and drag it to form a new $\triangle ABC$. What is true about the ratios $\dfrac{BM}{BA}$ and $\dfrac{BN}{BC}$?

    b. Drag M to a new position. What is true about the ratios?

6. a. Measure MN and AC. How does the ratio $\dfrac{MN}{AC}$ compare to the ratios $\dfrac{BM}{BA}$ and $\dfrac{BN}{BC}$?

    b. Click on B and drag it to form a new triangle. Now how do the ratios in Part a compare?

7. What seems to be true when a line segment intersects two sides of a triangle and is parallel to the third side?

8. What can you conclude about $\triangle ABC$ and $\triangle MBN$? Justify your answer.

1. a. Construct △*ABC*.

   b. Locate *D, E,* and *F,* the midpoints of $\overline{AB}$, $\overline{BC}$, and $\overline{AC}$, respectively.

   c. Draw the three segments connecting *D, E,* and *F.*

2. a. Locate *G, H, I, J, K,* and *L,* the midpoints of $\overline{DB}$, $\overline{BE}$, $\overline{EC}$, $\overline{CF}$, $\overline{FA}$, and $\overline{AD}$, respectively.

   b. Draw segments $\overline{GH}$, $\overline{GJ}$, $\overline{HK}$, $\overline{LI}$, $\overline{LK}$, and $\overline{IJ}$.

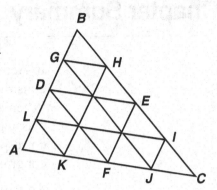

3. Use the Measure Menu to find the following ratios.

   a. $\dfrac{DE}{AC}$ $\dfrac{\text{perimeter of } \triangle DBE}{\text{perimeter of } \triangle ABC}$ $\dfrac{\text{area of } \triangle DBE}{\text{area of } \triangle ABC}$

   b. $\dfrac{HK}{AB}$ $\dfrac{\text{perimeter of } \triangle KHC}{\text{perimeter of } \triangle ABC}$ $\dfrac{\text{area of } \triangle KHC}{\text{area of } \triangle ABC}$

   c. $\dfrac{DF}{GJ}$ $\dfrac{\text{perimeter of } \triangle ADF}{\text{perimeter of } \triangle AGJ}$ $\dfrac{\text{area of } \triangle ADF}{\text{area of } \triangle AGJ}$

4. How are the ratios in each part of Exercise 3 related?

5. a. Click on point *B* and drag it to form a new △*ABC*.

   b. Repeat Exercise 3.

   c. Compare the values of the ratios in Exercise 3 with those in Part b. What did you find?

6. a. List three pairs of non-congruent similar triangles in the figure above.

   b. Explain how you determined that the triangles are similar.

   c. What is the ratio of the lengths of the corresponding sides in each pair?

   d. What is the ratio of the perimeters of each pair?

   e. What is the ratio of the areas of each pair?

7. List two pairs of congruent triangles. Explain how you determined that the triangles are congruent.

Chapter Summary

In Activity 1, the *Triangle Inequality* was reinforced as you constructed triangles with different combinations of side lengths. When a triangle could be constructed, comparing your triangle with a classmate's introduced the idea that triangles are congruent if their corresponding sides are congruent. This is the Side-Side-Side (SSS) congruence property of triangles.

Activity 2 developed the Side-Angle-Side (SAS), Angle-Side-Angle (ASA), and Angle-Angle-Side (AAS) properties for congruence of triangles. When given the measures of two sides and an angle that is not included by them (SSA), you discovered that, depending on the given measures, there were three possibilities: exactly one triangle could be constructed, two distinct triangles could be constructed, or no triangle could be constructed.

In Activity 3, pattern blocks were used to introduce the concept of similarity. Using the blocks, you could easily determine that in similar figures the corresponding angles are congruent and the ratios of the lengths of corresponding sides are equal. You also discovered that two polygons could have either corresponding angles congruent or the ratios of corresponding sides equal but not be similar. Both conditions are necessary for the polygons to be similar.

In Activity 4, when you constructed a triangle given the measures of the angles, you discovered through comparison of your triangle with a classmate's, that the triangles were similar.

Geometry surrounds us in our world and applications of geometry abound. *The Outdoor Geometry* activity allowed you to apply the concepts developed in earlier activities in real-world problem settings.

In Activity 6, you formed a new triangle by constructing a segment parallel to a side in a given triangle. Then you investigated the relationships between the sides of the two triangles, between their perimeters, and between their areas.

Chapter 13
Statistics: The Interpretation of Data

"The amount of data available to help make decisions in business, politics, research, and everyday life is staggering: Consumer surveys guide the development and marketing of products. Polls help determine political-campaign strategies, and experiments are used to evaluate the safety and efficacy of new medical treatments. Statistics are often misused to sway public opinion on issues or to misrepresent the quality and effectiveness of commercial products. Students need to know about data analysis and related aspects of probability in order to reason statistically—skills necessary to becoming informed citizens and intelligent consumers."

—Principles and Standards for School Mathematics

As noted in the *Principles and Standards*, the public is inundated with data of all sorts presented in a variety of formats. A wise consumer must be able to analyze data and the manner in which it is presented in order to critically evaluate the claims of advertisers, politicians, and others. One must also learn to judge the reasonableness of conclusions and/or predictions that are derived from data.

Statistics may be defined as the science of collecting, organizing, and interpreting data. In this chapter, you will learn to collect and sample data and to organize them using a variety of techniques—dot plots, frequency tables, bar graphs, stem-and-leaf plots, and box-and-whisker plots. You also will learn to describe data using measures of central tendency and to interpret data presented in graphs.

The activities in this chapter are designed to develop your understanding of these concepts and the processes of statistics. The emphasis in the activities is on the visual presentation of data and informal methods of data analysis rather than on formal statistical methods. In the activities, you will learn to use statistics to communicate information about sets of data effectively. You will also learn to interpret statistical displays and to make critical and informed decisions based on them.

Correlation of Chapter 13 Activities to the
Common Core Standards of Mathematical Practice

Activity Number and Title		Standards of Mathematical Practice
1:	Graphing *m&m's*®	SMP 3, SMP 4, SMP 5, SMP 7
2:	Grouped Data	SMP 2, SMP 3, SMP 4, SMP 7
3:	What's the Average?	SMP 2, SMP 3, SMP 4, SMP 5, SMP 7
4:	Finger-Snapping Time	SMP 2, SMP 3, SMP 4, SMP 5, SMP 7
5:	The Weather Report	SMP 1, SMP 2, SMP 3, SMP 4, SMP 5, SMP 7
6:	Populations and Samples	SMP 1, SMP 2, SMP 3, SMP 4, SMP 5, SMP 7

Activity 1: Graphing *m&m's*®

PURPOSE	Use a variety of graphs to display data and explore relationships among the data.
COMMON CORE SMP	SMP 3, SMP 4, SMP 5, SMP 7
MATERIALS	Online: Six *m&m's*® Line Plot charts
	Other: A one-pound bag of *m&m's*® Milk Chocolate Candies*, a one-tablespoon measure, colored pencils, a balance scale and weights, and a calculator
GROUPING	Work individually and as a whole class.
GETTING STARTED	*m&m's* Milk Chocolate Candies come in six colors: brown, green, orange, red, blue, and yellow.

Before you take a sample from the bag of *m&m's*, answer the following.

1. a. Which color of *m&m's* do you think will occur most often in the bag? Why?

   b. in your sample? Why?

2. a. Which color of *m&m's* do you think will occur least often in the bag? Why?

   b. in your sample? Why?

REAL GRAPHS

1. Take a sample of *m&m's* by dipping the measuring spoon into the bag of candy and removing a spoonful. *CAUTION: Do not eat any of the m&m's!*

Statistical data are often displayed graphically. Using a graph rather than simply presenting the data as a set of numbers makes it easier to study relationships in the data.

2. Arrange the *m&m's* in your sample on Graph 1 on the next page. This type of graph is often called a *real graph* because the statistical data are displayed using the actual objects whose frequencies are being compared.

3. Record the number of *m&m's* of each color and the total number in your sample.

   Brown: ____ Orange: ____ Blue: ____ Green: ____ Red: ____ Yellow: ____ Total: ____

4. What color occurred most often and what color occurred least often in your sample? How do these colors compare with your predictions?

---

* *m&m's* Milk Chocolate Candies is a registered trademark of Mars, Inc.

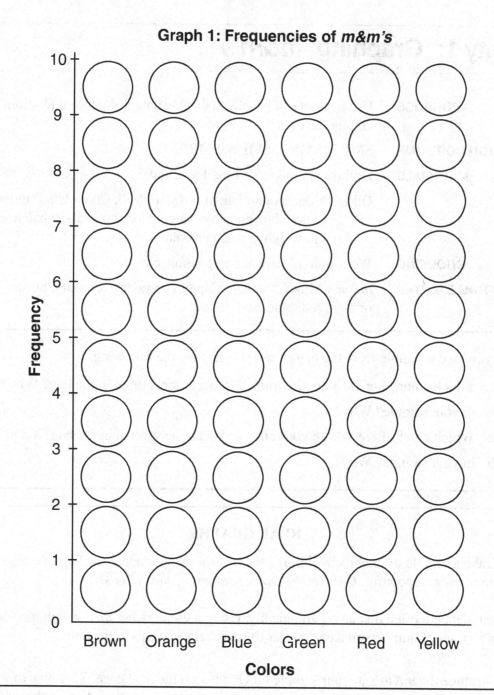

Graph 1: Frequencies of *m&m's*

PICTOGRAPHS

1. As you remove each candy from the graph, color its circle the appropriate color. This type of graph is called a *pictograph* because the data are displayed using parallel columns (or rows) of pictures in which each picture represents one or more of the objects being compared.

Now you may eat the m&m's in your sample!

2. Compare your pictograph with your classmates' pictographs. Describe any similarities and differences and explain why these may have occurred.

DOT PLOTS

1. Complete the following to collect the class data for the yellow *m&m's.*

    a. What is the maximum number of yellow *m&m's* in anyone's sample?

    b. What is the minimum number of yellow *m&m's* in anyone's sample?

    c. Title one of the line-plot charts "Yellow" and use the maximum and minimum values from Parts a and b to label the scale on the number line.

    d. Each time a person reports the number of yellow *m&m's* in his or her sample, record an X above that number on the number line.

    Example:

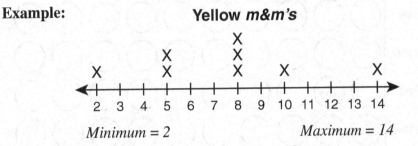
Yellow *m&m's*

    Minimum = 2   *Maximum = 14*

This type of graph is called a *dot* or *line plot.* Dot plots provide a quick, simple way to organize numerical data. They work best when there are fewer than 25 data points.

2. Repeat Exercise 1 for each color of *m&m's.*

3. Use the dot plots to describe the data for each color. Rather than just looking at individual numbers, describe the shape of the data—any patterns or special features such as clusters or gaps in the data and isolated data points—that tell how the data are distributed.

4. Use the dot plots to find the total number of each color of *m&m's* in the samples.

    Brown: ____ Orange: ____ Blue: ____ Green: ____ Red: ____ Yellow: ____

PREDICTIONS

1. Use the class data to predict the number of each color of *m&m's* that you would expect to find in a one-pound bag.

    Brown: ____ Orange: ____ Blue: ____ Green: ____ Red: ____ Yellow: ____

2. Describe the procedure you used to make your predictions.

3. Help your classmates count the *m&m's* remaining in the bag. Add these counts to the numbers you already have. What was the total number of each color of *m&m's* in the bag?

    Brown: ____ Orange: ____ Blue: ____ Green: ____ Red: ____ Yellow: ____

4. How do these totals compare with the predictions you made in Exercise 1?

PICTOGRAPHS REVISITED

Construct a pictograph for the number of each color of *m&m's* in the bag on Graph 2 below.
HINT: Let each circle represent more than one *m&m*.

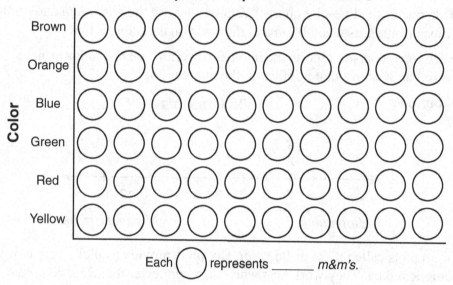

Graph 2: Frequencies of *m&m's*

Each ( ) represents _____ *m&m's*.

BAR GRAPHS

1. Use Graph 3 below to construct a horizontal bar graph for the number of each color of *m&m's* in the bag. Label the scale on the horizontal axis.

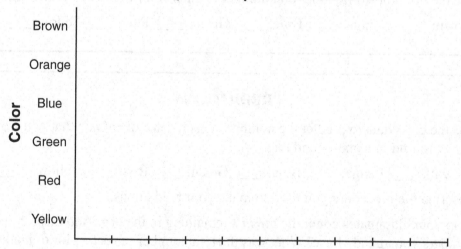

Graph 3: Frequencies of *m&m's*

2. a. Which graph, the pictograph or the bar graph, was easier to construct? Why?

    b. Which graph is easier to read? Why?

Activity 2: Grouped Data

PURPOSE	Display data using grouped frequency tables, histograms, and stem-and-leaf plots.
COMMON CORE SMP	SMP 2, SMP 3, SMP 4, SMP 7
MATERIALS	Other: Protractor
GROUPING	Work individually or in pairs.
GETTING STARTED	In many situations, there may be so much data, or the data may be so spread out, that it becomes difficult to construct a dot plot or a frequency table for individual items. In such cases, it may be more convenient to group the data.

President	Political Party[1]	Age at Inauguration	President	Political Party[1]	Age at Inauguration
Washington	F	57	B. Harrison	R	55
J. Adams	F	61	Cleveland	D	55
Jefferson	DR	57	McKinley	R	54
Madison	DR	57	T. Roosevelt	R	42
Monroe	DR	58	Taft	R	51
J. Q. Adams	DR	57	Wilson	D	56
Jackson	D	61	Harding	R	55
Van Buren	D	54	Coolidge	R	51
W. H. Harrison	W	68	Hoover	R	54
Tyler	W	51	F. D. Roosevelt	D	51
Polk	D	49	Truman	D	60
Taylor	W	64	Eisenhower	R	62
Fillmore	W	50	Kennedy	D	43
Pierce	D	48	L. B. Johnson	D	55
Buchanan	D	65	Nixon	R	56
Lincoln	R	52	Ford	R	61
A. Johnson	U	56	Carter	D	52
Grant	R	46	Reagan	R	69
Hayes	R	54	G. H. W. Bush	R	64
Garfield	R	49	Clinton	D	46
Arthur	R	51	G. W. Bush	R	54
Cleveland	D	47	Obama	D	47

[1]F = Federalist, DR = Democratic-Republican, D = Democrat, W = Whig, R = Republican, U = Unionist
SOURCE: The World Almanac and Book of Facts 2011.

GROUPED FREQUENCY TABLES

1. Complete the grouped frequency table for the presidents' ages at inauguration.

2. How many different ages can fall within an interval in the table?

3. Can you determine the range, median, and mode of the ages from the information in the table? Explain why or why not for each statistic.

Presidents' Ages at Inauguration		
Interval	**Tally**	**Frequency**
40–44		
45–49		
50–54		
55–59		
60–64		
65–69		

HISTOGRAMS

The data in the grouped frequency table can be pictured graphically using a *histogram,* a graph closely related to a bar graph.

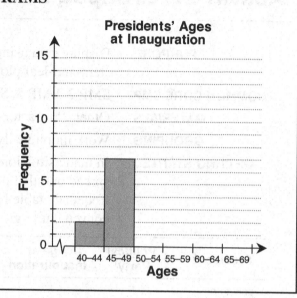

Presidents' Ages at Inauguration

1. How are the bars on the histogram related to the frequencies in the table?

2. Complete the histogram.

3. How is a histogram like a bar graph? How is it different?

4. Can you tell from the histogram how many presidents were 57 years old when they were inaugurated? Explain.

ORDERED STEM-AND-LEAF PLOTS

Stem-and-leaf plots provide a natural way to group data in intervals. To construct a stem-and-leaf plot, first find the least and the greatest data points.

What is the youngest age at inauguration for the presidents? _____ the oldest age? _____

Next, decide on the stems. Since the presidents were inaugurated in their 40s, 50s, and 60s, use the tens digits of the ages as the stems. Write the stem digits in a column from least to greatest on the left side of a vertical line, as shown.

Stem	Leaf
4	
5	
6	

For each age at inauguration, record a leaf by writing the units digit of the age on the right side of the vertical line in the row that contains its stem. The leaves for Washington and J. Adams have been done for you.

Stem	Leaf
4	
5	7
6	1

After all the leaves have been recorded, rearrange the leaves in increasing order and add a title and a key to the plot.

1. Use the stem-and-leaf plot to find the range, median, and mode of the ages at inauguration.

2. How are a stem-and-leaf plot and a histogram alike? How do they differ?

Presidents' Ages at Inauguration

Stem	Leaf
4	
5	
6	

5 | 7 = 57 years old

BACK-TO-BACK STEM-AND-LEAF PLOTS

1. On the plot, list the stems for the age at inauguration in the center column. Construct a stem-and-leaf plot for the age at inauguration of the Democratic (D) presidents to the left of the stems and one for the Republican (R) presidents to the right of the stems.

2. How do the ages at inauguration of Democratic presidents compare to those of Republican presidents?

Presidents' Ages at Inauguration

Democratic		Republican

EXTENSIONS

A group of adults, all at least 30 years old, were asked, "How many hours of sleep do you usually get a night?" The results of the survey are summarized in the histogram and circle graph at the left.

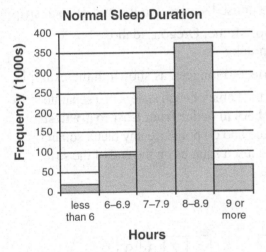

Normal Sleep Duration

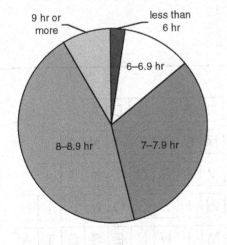

Normal Sleep Duration

1. a. The number of people who sleep 8–8.9 hr a night is about how many times the number who sleep 6–6.9 hr a night?

   b. Does the circle graph or the histogram make it easier to find this information? Why?

2. a. About what fraction of the people sleep 8–8.9 hr a night?

   b. Which graph did you use?

3. a. Use the histogram to estimate the number of people surveyed. Explain how you made your estimate.

   b. Use your estimate from Part a to find about what percent of the people sleep 7–7.9 hr each night.

4. a. Use a protractor to measure the central angle for the 7–7.9 hr sleep category. What percent of 360° is this?

   b. How does your answer in Part a compare with your answer in Exercise 3 Part b?

Activity 3: What's the Average?

PURPOSE   Investigate mean, median, and mode and examine how each average is affected by extremes in the data.

COMMON CORE SMP   SMP 2, SMP 3, SMP 4, SMP 5, SMP 7

MATERIALS   Online: Centimeter Graph Paper
Other: Scissors

GROUPING   Work individually or in groups of 3 or 4.

GETTING STARTED   Cut out twenty 1 cm × 17 cm strips of graph paper.

When we describe a set of data, it is often convenient to use a single number, often called the *average*, to indicate where the data are centered or concentrated. The mean, the median, and the mode are three commonly used *averages*.

Write the name of each of the following states on a strip of graph paper. Use one strip of paper for each state and one square for each letter in the name. Cut off the unused squares on the end of each strip.

Arizona, Hawaii, Ohio, Maine, Oregon, Idaho, Texas, Louisiana, Kentucky

Arrange the names from shortest to longest, as shown in the example.

Count the number of letters in the name of each state. On a separate strip of graph paper, **write the numbers in order from least to greatest**. Write one number in each square and do not leave any blank squares between numbers. Cut off the unused squares on the end of the strip.

Example:

N	E	V	A	D	A

M	O	N	T	A	N	A

V	I	R	G	I	N	I	A		6	7	8	9	9

W	I	S	C	O	N	S	I	N

M	I	N	N	E	S	O	T	A

THE MODE

Look at the numbers on the strip. What number of letters occurs most often in the names of the states?

This is the **mode** of the numbers of letters in the names of the states.

In the example, the mode is 9.

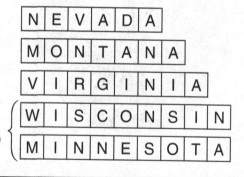

Mode (9 letters)

THE MEDIAN

1. Fold the strip containing the numbers of letters in the names of the states in half by folding the ends together.

2. Unfold the strip. Through which number does the fold pass?

This is the **median** number of letters in the names of the states. In the example, the median is 8.

3. If the fold is on the line between two numbers, what number would you use for the median? Why?

4. How many states have names that contain fewer letters than the median? more letters than the median?

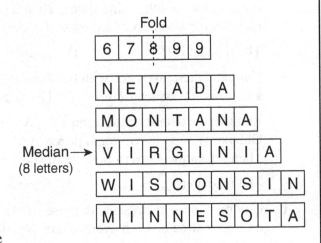

THE MEAN

To find the **mean** of the numbers of letters in the names, cut off letters from the longer names and move them to fill in the shorter ones. Continue cutting off and moving letters until all the rows have as close to the same number of letters as possible.

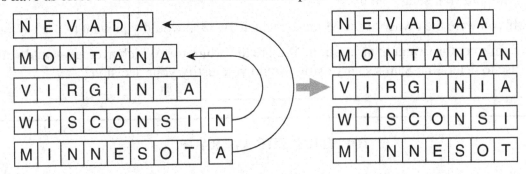

The mean in this example is a little less than eight because all the rows except one contain eight letters.

What is the mean of the numbers of letters in the names of the nine states?

1. Write each letter of Massachusetts in a square on a strip of graph paper. Add this to the data for the other nine states. Then repeat the steps to find the median, mode, and mean of the numbers of letters in the names of the ten states.

    The median is _____.   The mode is _____.   The mean is _____.

2. Compare these averages with those for the original nine states. Describe how the addition of Massachusetts affected each average and explain the differences.

3. Remove the data for Massachusetts. Write the names Maryland, Michigan, and Oklahoma on strips of graph paper. Add them to the data for the original nine states. Then repeat the steps to find the median, mode, and mean.

    The median is _____.   The mode is _____.   The mean is _____.

4. Compare these averages with those for the original nine states. Describe how these additions affected each average and explain the differences.

5. To find the mean of the numbers of letters in the names of N states, the letters making up the names of the states must be separated into N sets with the same (or nearly the same) number of letters in each set. Explain how you could find the number of letters in each set without writing each letter on a square.

WHICH WOULD YOU USE?

Sam Slugger's contract with the Columbus Mudcats baseball team says his annual salary will be $1,000,000 times the *average* of his batting averages for the preceding five seasons. Sam's batting averages for the past five seasons were .145, .130, .160, .130, and .495.

1. If you were Sam, which *average*—mean, median, or mode—would you want to use to compute your salary? Why?

2. If you were the owner of the Mudcats, which *average* would you want to use? Why?

3. Sam's contract went to arbitration. You are the arbitrator. Which average would you use to determine Sam's salary? How would you justify your decision?

IDENTIFY THE AVERAGE

Which *average*—mean, median, or mode—do you think was used in each of the following statements? Explain your choice in each case.

The *average* lady's shoe size is $7\frac{1}{2}$.

The *average* size of a household in the United States is 2.62 people.

The *average* household income in the United States is $52,029.

Activity 4: Finger-Snapping Time

PURPOSE	Display and compare data using box-and-whisker plots.
COMMON CORE SMP	SMP 2, SMP 3, SMP 4, SMP 5, SMP 7
MATERIALS	Online: Centimeter Graph Paper Other: Scissors and a clock or watch to measure elapsed time in seconds
GROUPING	Divide the class into groups of 10 to 12 students each; students work in pairs.
GETTING STARTED	Cut out three 1 cm × 17 cm strips of graph paper.

Snap your fingers as fast as you can for 15 seconds. Have your partner time you while you snap your fingers and count the number of snaps. Then do the same thing for your partner. Record the data in the table below.

Ask the other people in your group how many times they snapped their fingers in 15 seconds and record the information in the table.

Finger Snaps I

Person	Finger Snaps in 15 sec	Person	Finger Snaps in 15 sec	Person	Finger Snaps in 15 sec
You		3		7	
Partner		4		8	
1		5		9	
2		6		10	

Box-and-whisker plots provide a useful method for summarizing and comparing data such as the number of times people can snap their fingers.

- **Step 1 in constructing a box-and-whisker plot is to order the data values in your group from least to greatest.**

Example:

20	35	37	39	41	45	45	47	59

1. Write the finger-snapping data in order from least to greatest on a strip of graph paper. Write one number in each square. Do not leave any blank squares between numbers. Cut off the unused squares on the end of the strip.

LE = 20; UE = 59

- **Step 2—find the extremes of the data.**

The least data point is called the *lower extreme* (LE), and the greatest data point is called the *upper extreme* (UE).

2. Find and record the lower extreme and the upper extreme of the finger-snapping data.

LE = _____   UE = _____

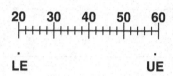

LE   UE

3.   Use the extremes to select an appropriate scale and label the number line below.

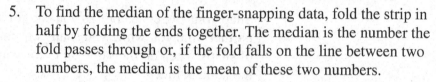

4.   Locate the lower and upper extremes by marking a dot under their coordinates on the scale, as in the example.

• **Step 3—find the median of the data.**

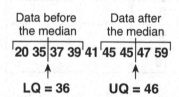

Median = 41

5.   To find the median of the finger-snapping data, fold the strip in half by folding the ends together. The median is the number the fold passes through or, if the fold falls on the line between two numbers, the median is the mean of these two numbers.

6.   Record the median and mark its location by making a dot under its coordinate on the scale.

Median = _____

• **Step 4—find the quartiles for the data.**

The *lower quartile* (LQ) is the median of the data values that occur before the median in an ordered list.

Data before   Data after
the median   the median

20 35│37 39│41│45 45│47 59

LQ = 36   UQ = 46

7.   Look at the part of the strip that contains data values that come before the median. Find the median of these values by folding this part of the strip in half. This is the lower quartile of the data.

Lower Quartile = _____

The *upper quartile* (UQ) is the median of the data values that occur after the median in an ordered list.

8.   Look at the part of the strip that contains data values that come after the median. Find the median of these values by folding this part of the strip in half. This is the upper quartile of the data.

Upper Quartile = _____

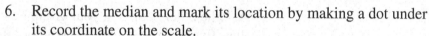

LQ   UQ

9.   Mark the locations of the upper and lower quartiles by making a dot for each below its coordinate on the scale.

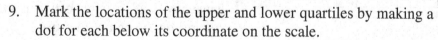

10. Form a box by drawing vertical segments through the dots for the upper and lower quartiles and connecting the endpoints of the segments, as in the example. Then draw a vertical segment through the median as shown.

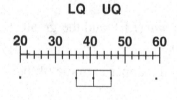

• **The final step is to identify and plot any outliers in the data.**

IQR = 46 − 36 = 10

11. Find the difference between the upper and lower quartiles. This difference is the *interquartile range* (IQR).

    Interquartile Range = UQ − LQ = _____

    If a data point is more than 1.5 interquartile ranges greater than the upper quartile or more than 1.5 interquartile ranges less than the lower quartile, it is an *outlier*.

    In the example, LQ − 1.5 × IQR = 36 − (1.5 × 10)

    = 21.

    Thus 20 is an outlier, since 20 < LQ − 1.5 × IQR.

12. Identify any outliers in your data. Mark their locations by making a dot for each one below its coordinate on the scale.

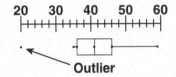

13. Complete the plot by drawing segments from the least data point that is not an outlier to the lower quartile and from the upper quartile to the greatest data point that is not an outlier. These segments are the whiskers.

14. Study the completed box-and-whisker plot. About what percent of the data points lie between the

    a. lower extreme and the lower quartile?

    b. upper quartile and the upper extreme?

    c. lower quartile and the median?

    d. median and the upper quartile?

15. About what percent of the data points lie in the box?

16. Repeat the finger-snapping experiment, only this time snap your fingers as fast as you can for 30 seconds. Divide the number of snaps by 2 to get your rate per 15 seconds Record your rate, your partner's rate, and the rates for the other people in your group in the table below.

Finger Snaps II

Person	Sex	Finger Snaps in 15 sec	Person	Sex	Finger Snaps in 15 sec	Person	Sex	Finger Snaps in 15 sec
You			3			7		
Partner			4			8		
1			5			9		
2			6			10		

17. a. Construct a box-and-whisker plot for the data in the table
       Finger Snaps II.

   b. In the space below the plot constructed in Part a, re-construct
       a box-and-whisker plot using the data for Finger Snaps I and
       the scale in Part a.

18. a. How do the medians of the two sets of data compare?

   b. The extremes?

   c. The interquartile ranges?

   d. The upper quartiles?

   e. The lower quartiles?

19. Based on your observations in Exercise 18, how do the finger-
    snapping rates in the first experiment compare to the rates in the
    second experiment? How would you explain any similarities or
    differences?

EXTENSIONS   1. a. Separate the data in the Finger Snaps II table into rates for
                       males and rates for females. Construct box-and-whisker plots
                       for the male and female rates using the same scale for both.

                 b. Based on your graphs, do you think there is any difference
                       between the finger-snapping rates for males and for females?
                       Explain your answers.

              2. a. Construct a stem-and-leaf plot for the data for Finger Snaps I.

                 b. What can you learn about the distribution of the data (gaps,
                       clusters, extremes, outliers, etc.) and the averages **from both**
                       the box-and-whisker plot and the stem-and-leaf plot of
                       the data?

                 c. What information can you get about the distribution of the
                       data and the averages from a box-and-whisker plot **but not
                       from** a stem-and-leaf plot?

                 d. What information can you get about the distribution of the
                       data and the averages from a stem-and-leaf plot **but not from**
                       a box-and-whisker plot?

Activity 5: The Weather Report

PURPOSE	Apply the concepts of data analysis, evaluate statements based on data, and analyze the advantages and disadvantages of using different displays of data.
COMMON CORE SMP	SMP 1, SMP 2, SMP 3, SMP 4, SMP 5, SMP 7
MATERIALS	Online: Half-centimeter Graph Paper Other: A calculator and an atlas
GROUPING	Work individually or in pairs.

Normal Daily Mean Temperature (°F) for Some U.S. Cities

Based on the Period 1971–2000

	Jan	Feb	Mar	Apr	May	Jun	July	Aug	Sep	Oct	Nov	Dec
Los Angeles, CA	57	58	58	61	63	66	69	71	70	67	62	58
San Francisco, CA	49	52	54	56	59	61	63	64	64	61	55	50
Wichita, KS	30	36	46	55	65	76	81	80	71	59	44	34
Portland, ME	22	25	34	44	54	63	69	67	59	48	38	28
Portland, OR	40	43	47	51	57	63	68	69	64	54	46	40
Charlotte, NC	42	45	53	61	69	77	80	79	73	62	52	44
Seattle, WA	42	44	47	51	57	61	66	66	61	53	46	41
Spokane, WA	27	33	40	47	54	62	69	69	59	47	35	27

Source: National Climatic Data Center

---

1. Construct a stem-and-leaf plot for the weather data given for San Francisco and Wichita.

San Francisco		Wichita
4		**3**
5		**4**
6		**5**
		6
		7
		8

2. Compute the mean, median, and mode for the data given for San Francisco and Wichita.

	San Francisco	Wichita
Mean		
Median		
Mode		

3. Construct line graphs for the temperatures for San Francisco and Wichita on the grid below.

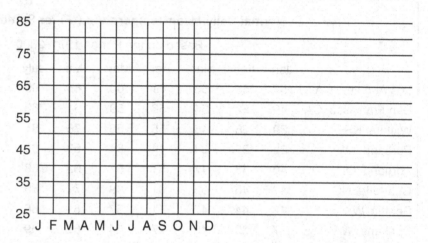

4. Using the scale on the grid above, construct vertical box-and-whisker plots for the temperatures for San Francisco and Wichita to the right of the line graphs.

5. Draw a horizontal line through the graphs to show the median for each set of data.

6. For how many months of the year is the temperature in Wichita within the range of the temperatures in San Francisco?

7. Which display of data can most easily be used to answer Exercise 6? Explain why or why not for each display.

8. The Wichita Chamber of Commerce could advertise that "the average annual temperature in Wichita is the same as that in *balmy* San Francisco." Evaluate this claim on the basis of the actual temperature data.

9. The mean and median annual temperatures for these two cities are nearly the same. Compare how accurately the mean and the median describe the climate in each city.

10. What other information do you need in addition to the medians or means to accurately describe the annual temperatures?

11. Find the latitude of Wichita: ____; and of San Francisco: ____. Which city is farther north?

12. If the difference in latitude is not significant, explain how the geographical location of each city affects its annual temperature.

13. Construct a back-to-back stem-and-leaf plot for Portland, ME, and Portland, OR.

Portland, ME		Portland, OR
	2	
	3	
	4	
	5	
	6	

14. On the grid below, construct line graphs and accompanying vertical box-and-whisker plots for the weather data for the two cities. Then draw the median lines. Explore questions similar to Exercises 6–12 for these two sets of data.

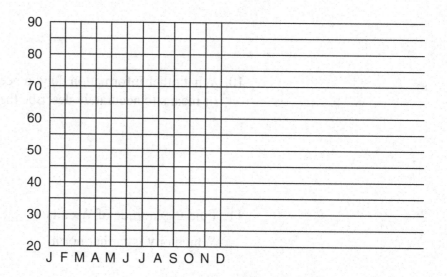

15. Choose other pairs of cities from the table and construct various displays for the weather data. Explore similarities and differences in the data. Determine the latitude and geographical location of each city. Explain how these affect the similarities or differences in the climates of the chosen cities.

Activity 6: Populations and Samples

PURPOSE	Simulate sampling techniques and explore the effect of sample size on the accuracy of predictions.
COMMON CORE SMP	SMP 1, SMP 2, SMP 3, SMP 4, SMP 5, SMP 7
MATERIALS	Other: Paper clip, beans, paper sandwich bag, and a colored marking pen
GROUPING	Work individually or in pairs.
GETTING STARTED	Statisticians often gather information and make predictions about a group of people or things called the *population*. Since it is often impractical or impossible to check every member of the population, a smaller group or *sample* is often studied.
	The table on the next page gives the amount spent per student by each state during the 2003–2004 school year. One way to select a random sample of the states is to use the spinner below the table to generate pairs of digits corresponding to the two-digit state numbers. If a number greater than 50 or a number already obtained is generated, ignore it and continue generating numbers until the desired number of states has been selected.

1. Use the spinner or the random integer generator on your calculator to select a random sample of 15 states. For each state in the sample, record the per student expenditure from the table.

2. Make a box-and-whisker plot of the expenditures for the states in the sample.

Per Student Expenditures for K–12 Public Schools

5,000   6,000   7,000   8,000   9,000   10,000   11,000   12,000

3. Based on the sample, what would you conclude about the per student expenditures for public K–12 schools in the United States?

4. The extremes, quartiles, and median of the per student expenditures for the 50 states are given below. Use them to make a box-and-whisker plot for the entire population below the one for the sample.

Extremes	
upper	$12,059
lower	$5,091

Quartiles	
upper	$9,169
lower	$7,163

Median
$7,623.50

5. Compare the box-and-whisker plots. How well do the characteristics of the sample reflect the characteristics of the population? Explain.

Per Student Expenditures for
Public K-12 Schools 2003–2004 ($)

01	Alabama	7,163	26	Montana	7,688
02	Alaska	9,808	27	Nebraska	7,352
03	Arizona	5,347	28	Nevada	6,230
04	Arkansas	6,005	29	New Hampshire	8,915
05	California	7,692	30	New Jersey	11,390
06	Colorado	8,023	31	New Mexico	7,370
07	Connecticut	11,773	32	New York	12,059
08	Delaware	10,470	33	North Carolina	6,727
09	Florida	6,516	34	North Dakota	6,835
10	Georgia	8,703	35	Ohio	9,136
11	Hawaii	8,220	36	Oklahoma	6,429
12	Idaho	6,372	37	Oregon	7,587
13	Illinois	9,839	38	Pennsylvania	8,609
14	Indiana	8,414	39	Rhode Island	10,258
15	Iowa	7,098	40	South Carolina	7,559
16	Kansas	7,622	41	South Dakota	7,300
17	Kentucky	7,474	42	Tennessee	6,279
18	Louisiana	7,179	43	Texas	7,335
19	Maine	10,145	44	Utah	5,091
20	Maryland	9,186	45	Vermont	10,630
21	Massachusetts	10,772	46	Virginia	6,441
22	Michigan	8,671	47	Washington	7,446
23	Minnesota	8,821	48	West Virginia	9,169
24	Mississippi	6,137	49	Wisconsin	9,483
25	Missouri	6,947	50	Wyoming	9,756

Source: NEA Rankings and Estimates Update—A Report of School Statistics—(Fall 2004)

Make a spinner by bending a paper clip into the shape shown below. The long, straight part will be the pointer. Place your pencil through the loop of the paper clip and put the point of the pencil on the center of the spinner. Spin the spinner by flicking the paper clip with your finger.

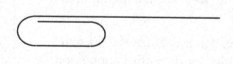

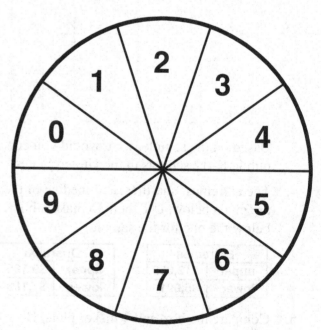

In the previous activity, you selected a random sample of 15 states. Could you make accurate predictions about the population by selecting smaller samples?

1. Select three random samples of 5 states and find the median per pupil expenditure for each sample. How do the medians of your three samples compare with the population median?

2. Select three random samples of 10 states and find the median per pupil expenditure for each sample. How do the medians of your three samples compare with the population median?

3. Were you able to accurately predict the median per pupil expenditure for all 50 states by analyzing samples of 5 or 10 states? Explain.

EXTENSION

1. Wildlife biologists use a sampling technique called *tag and recapture* to estimate a population's size. Complete the following to simulate this method.

   a. Pour a bag of beans into the paper bag. Each bean represents one animal.

   b. Take out a handful of beans and mark them. Record how many beans you marked.

   c. Put the marked beans back in the bag and mix the beans thoroughly.

   d. Take out another handful of beans. Count the number of beans and how many are marked.

2. The beans you removed in Exercise 1 Part d are a sample of the animal population.

   a. What is the sample size?

   b. Write a ratio of the number of marked beans in your sample to the sample size.

3. Let P = the total number of beans in the bag (the size of the animal population). Write a ratio of the number of marked beans in the bag to the total number of beans in the bag.

4. a. What should be true about the ratios you wrote in Exercises 2 and 3? Why?

   b. Use this fact to estimate P.

5. Why is this sampling technique called *tag and recapture*?

Chapter Summary

Activities 1, 2, and 4 introduced a variety of graphs, tables, and plots that can be used to organize and display data. The focus of the activities was on the potential uses and limitations of each statistical display, not just the construction of each display.

In general, *real graphs* do not provide a practical method for displaying statistical data. *Pictographs* are common in the popular media, probably because of their aesthetic appeal, but they are difficult to interpret and construct. *Bar graphs*, on the other hand, display exactly the same information and have the advantage of being easy to construct and to interpret.

Because *dot plots*, or *line plots*, are quick and easy to construct, they provide a very useful tool for preliminary data analysis. Maximum and minimum values, clusters and gaps in the data, outliers, the median, and the mode are all easily identified on a dot plot. The major limitation is that dot plots are convenient to use only with relatively small sets of data.

Stem-and-leaf plots and *histograms* provide information about how data are distributed. Maximum and minimum values, the median, and mode, clusters and gaps in the data, and outliers are all easily identified in a stem-and-leaf plot. Histograms show gaps and clusters just as stem-and-leaf plots do, but you cannot identify individual data points as you can in stem-and-leaf plots. *Box-and-whisker plots* focus on the range and distribution of the data. Clustering of the data can sometimes be identified, but gaps in the data cannot. Box-and-whisker plots are very useful when working with large sets of data and for comparing different sets of data.

Activity 3 introduced the concept of measures of central tendency, or averages, and developed the meanings of the mean, median, and mode. The development focused on (a) how each measure provides an indication of where data are centered or concentrated, (b) how each measure is affected by extremes, and (c) how the physical modeling is related to the algorithm for computing the mean.

The activity illustrated that the *mean* has a leveling or smoothing effect. Another way of expressing this is, if all the data had the same value, the data points would all equal the mean. The mean is the most commonly used average—in fact, many people erroneously use the terms *average* and *mean* synonymously. However, the value of the mean may be affected by extremes in the data, and therefore it may not be the most representative value to use for an average.

The *median* is the middle data point or the mean of the two middle data points. Unusually large or small data points do not significantly affect the value of the median; thus the median is a more appropriate average than the mean when there are extremes in the data. However, the median may not accurately reflect concentrations in the data.

The *mode*, the data point that occurs most often, is a measure of where the data are concentrated. Its usefulness is limited, however, since the frequency of occurrence of the mode may not be significantly different than that of other data points, the mode may be an outlier, or the data may have more than one mode.

Activities 1 and 6 explored the relationship between a sample and the population. Predicting the characteristics of a population from a sample is an important concept in statistics. The reliability of such predictions is affected by two factors: the sample size and the randomness of the sample. Different sampling techniques and the effect of sample size were explored in Activity 6. Assuming the samples are selected randomly, predictions generally become more accurate as the size of the sample increases. Randomness was demonstrated in Activity 1 by the fact that there was a great deal of variation in the individual samples of *m&m's*®.

The focus of Activity 5 was on analyzing various plots to determine what information could be more easily derived from one display than another and to evaluate the use of mean or median as a good descriptor of average.

Advertisers, policy makers, and decision-makers constantly use the term *average*, but the average reported may not accurately describe the entire set of data. It is very important to know the range of data to fully understand what the mean or median really indicates. For example, the mean and median temperatures for San Francisco and Wichita differ by only a degree. However, the range of the data varies considerably.

The stem-and-leaf plots illustrate the difference in the range of data, but the difference is much more visually apparent in line graphs or in side by side box-and-whisker plots. The median and the upper and lower quartiles of data can be identified in a stem-and-leaf plot; however, the relationships among these measures are much clearer in box-and-whisker plots. Each display has its own unique characteristics, and each offers a different insight into the data.

Chapter 14
Probability

"Teachers should build on children's developing vocabulary to introduce and highlight probability notions, for example, We'll *probably* have recess this afternoon, or It's *unlikely* to rain today. Young children can begin building an understanding of chance and randomness by doing experiments with concrete objects, such as choosing colored chips from a bag. In grades 3–5 students can consider ideas of chance through experiments—using coins, dice, or spinners—with known theoretical outcomes or through designating familiar events as impossible, unlikely, likely, or certain. Middle-grades students should learn and use appropriate terminology and should be able to compute probabilities for simple compound events, such as the number of expected occurrences of two heads when two coins are tossed 100 times. ... Through the grades, students should be able to move from situations for which the probability of an event can readily be determined to situations in which sampling and simulations help them quantify the likelihood of an uncertain outcome."

—Principles and Standards for School Mathematics

"Students develop a probability model and use it to find probabilities of events. Compare probabilities from a model to observed frequencies; if the agreement is not good, explain the possible sources of the discrepancy ... Students find probabilities of compound events using organized lists, tables, tree diagrams and simulation."

—Common Core State Standards for Mathematics

The activities in this chapter develop the basic ideas and vocabulary of probability and introduce several different models used to determine probabilities. You will learn to use ratios to assign a probability to an event and to use and interpret both experimental and theoretical probabilities. In the process, you will develop an understanding of fair and unfair games and random events.

Most people have an intuitive understanding of probability even though they have had limited formal educational experiences with it. The goal of this chapter is to extend your intuitive ideas to a sound mathematical understanding of probability.

Correlation of Chapter 14 Activities to the
Common Core Standards of Mathematical Practice

Activity Number and Title		Standards of Mathematical Practice
1:	What Are the Chances?	SMP 2, SMP 3, SMP 4, SMP 5
2:	The Spinner Game	SMP 2, SMP 3, SMP 4, SMP 5, SMP 7
3:	Theoretical Probability	SMP 2, SMP 3, SMP 4, SMP 5, SMP 6, SMP 7, SMP 8
4:	Paper-Scissors-Rock	SMP 2, SMP 3, SMP 4, SMP 5, SMP 7, SMP 8
5:	How Many Arrangements?	SMP 1, SMP 2, SMP 3, SMP 4, SMP 5, SMP 7, SMP 8
6:	Pascal's Probabilities	SMP 2, SMP 3, SMP 4, SMP 5, SMP 7
7:	Simulate It	SMP 1, SMP 2, SMP 3, SMP 4, SMP 5, SMP 7, SMP 8

Activity 1: What Are the Chances?

PURPOSE	Introduce the concept of probability using intuitive ideas about chance.
COMMON CORE SMP	SMP 2, SMP 3, SMP 4, SMP 5
GROUPING	Work individually.
GETTING STARTED	We frequently encounter situations in which we cannot predict the outcome in advance. In such situations, we often talk about the chances of an outcome occurring. If we think the chances that something will happen are good, we might say it is *likely* or it is probable. On the other hand, when we think the chances are poor, we often say the outcome is *unlikely* or not very probable.

A paper bag contains eight marbles: five green, two blue, and one yellow. Suppose one marble is drawn from the bag. Use the scale below to describe the chances of each of the following events occurring. Explain your decision for each event.

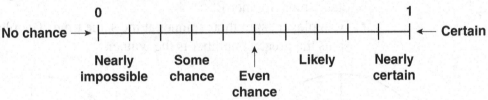

1. A green marble is drawn.   2. A blue marble is drawn.

3. A yellow marble is drawn.   4. A red marble is drawn.

5. A blue or yellow marble is drawn. 6. The marble drawn is not yellow.

Make a spinner face by dividing each circle into three sections and labeling each section with a color (RED, WHITE, or BLUE) so that the given condition is true.

1. The spinner is certain to stop on RED.

2. The spinner is likely to stop on WHITE.

3. The spinner can't stop on BLUE.

4. There is little chance the spinner will stop on RED.

5. The spinner will probably stop on RED or BLUE.

6. The spinner has the same chance of stopping on RED, WHITE, or BLUE.

Activity 2: The Spinner Game

PURPOSE	Introduce experimental probability, fair games, and mutually exclusive, complementary, certain, and impossible events.
COMMON CORE SMP	SMP 2, SMP 3, SMP 4, SMP 5, SMP 7
MATERIALS	Other: A paper clip for each student
GROUPING	Work in pairs.
GETTING STARTED	Make a spinner by bending a paper clip into the shape shown at the left. The long, straight part will be the pointer. Place the point of your pencil through the loop of the paper clip and put the point on the center of the spinner. Spin the spinner by flicking the paper clip with your finger.

Rules for the Spinner Game:

- This is a game for two players. One player spins Spinner A and the other spins Spinner B.
- Both players spin their spinner at the same time. The player who spins the greatest number is the winner.

Spinner A

Spinner B

1. Play the game 30 times. Record the results for each game in the tables below.

Number on Spinner A	Tally	Frequency
2		
4		
9		
Total		

Winning Spinner		
Spinner	Tally	Frequency
A		
B		
Total		

2. Do you think the Spinner Game is a fair game? Explain.

1. Combine your data for the winning spinner with the data from the other teams in your class to find a class total for the number of wins for Spinner A and for Spinner B. Record the data in the table.

Winning Spinner	
Spinner	**Frequency**
A	
B	
Total	

2. Does this data change your opinion about whether the game is fair or not? Explain.

EXPERIMENTAL PROBABILITY

A meteorologist says *the chance of snow today is 20%.* The fine print in the sweepstakes announcement indicates that *your odds of winning are 1 in 2.8 million.* A report states that *22 out of every 25 nineteen year-old women* will never have been married. These are just three examples of situations where probabilities are used to make predictions.

A *probability* is a ratio that describes the chance or likelihood of something happening.

$$\text{Experimental probability of an event} = \frac{\text{number of times the event occurs}}{\text{total number of trials}}$$

1. Use the data in the table for the numbers on Spinner A to calculate the following experimental probabilities. [$P(2)$ = the probability the spinner stops in the region labeled 2.]

   $P(2) = $ _____ $P(4) = $ _____ $P(9) = $ _____

   $P(\text{square number}) = $ _____ $P(\text{odd number}) = $ _____ $P(\text{even number}) = $ _____

2. What is $P(1)$? Why?

3. What is $P(\text{a number less than 10})$? Why?

4. What does it mean for an event to have

   a. a probability of 0?   b. a probability of 1?

5. Explain why the probability of an event can never be greater than 1 or less than 0.

COMPLEMENTARY EVENTS

1. What is P(even number) + P(odd number)?

2. If the spinner does not stop on an even number, then it must stop on an odd number. For this reason, the events "the spinner stops on an even number" and "the spinner stops on an odd number" are called complementary events. What can you conclude about the probabilities of complementary events?

MUTUALLY EXCLUSIVE EVENTS

1. a. What is P(2 or 4)?

    b. What is P(2 and 4)?

    c. The events "the spinner stops in the region labeled 2" and "the spinner stops in the region labeled 4" cannot happen at the same time. Events that cannot occur simultaneously are *mutually exclusive*. What is P(2) + P(4)?

    d. Based on your answers to Parts a–c, if A and B are mutually exclusive events, what is P(A or B)?

2. What is the difference between mutually exclusive events and complementary events?

3. a. What is P(an odd number or a square number)?

    b. What is P(an odd number) + P(a square number)?

    c. What is P(the number is odd and square)?

    d. Are the events "the spinner stops on an odd number" and "the spinner stops on a square number" mutually exclusive? Explain.

    e. How could you use your answers to Parts b and c to find P(an odd number or a square number)?

    f. If A and B are any two events, what is P(A or B)?

EXTENSION   Play the following spinner game.

Rules:
- This is a game for two players.
- The first player chooses a spinner from Spinners A, B (from the Spinner Game), or C (below).
- The second player chooses a spinner from the two remaining spinners.
- Both players spin at the same time. The player who spins the greatest number is the winner.

Spinner C

Play the game using different combinations of spinners. Then answer the following questions.

1. How does allowing the first player to choose a spinner affect the fairness of this game?

2. Is there a strategy for choosing the spinners that would give one player an advantage over the other? Explain.

Activity 3: Theoretical Probability

PURPOSE Introduce theoretical probability, equally likely events, and the use of a matrix to find experimental probabilities.

COMMON CORE SMP SMP 2, SMP 3, SMP 4, SMP 5, SMP 6, SMP 7, SMP 8

GROUPING Work individually or in pairs.

GETTING STARTED When you spin Spinner A, there are three possible outcomes. Since each central angle on the spinner has the same measure, over many trials, each of the outcomes should occur about the same number of times. That is, the outcomes are *equally likely*.

Spinner A

If the outcomes of an experiment are equally likely, the theoretical probability of an event may be calculated without conducting an experiment.

$$\text{Theoretical probability} = \frac{\text{number of outcomes making up the event}}{\text{total number of possible outcomes}}$$

---

1. Calculate the following theoretical probabilities for spinning Spinner A.
    [$P(2)$ = the theoretical probability the spinner stops in the region labeled 2.]

    $P(2) =$ _____ $P(4) =$ _____ $P(9) =$ _____

    $P(\text{square number}) =$ _____ $P(\text{odd number}) =$ _____ $P(\text{even number}) =$ _____

2. Compare the probabilities in Exercise 1 with the experimental probabilities you calculated in Exercise 1 of Activity 2. Explain any similarities or differences.

---

1. Design a spinner that has two regions, one labeled **Red** and the other **Blue**, and such that $P(\text{Red}) = \frac{1}{4}$.

2. Spin the spinner 20 times and record the number of times it stops on Red. Use the results to calculate the following experimental probabilities.

    $P(\text{Red}) =$ _____ $P(\text{Blue}) =$ _____

3. Did the probabilities come out exactly as you expected? Explain.

4. Combine the results of your 20 spins with those of four classmates. Calculate the experimental probabilities for the 100 spins. What do you observe?

THE SPINNER GAME REVISITED

1. Use your individual data from Activity 2 to calculate the following experimental probabilities.

    P(Spinner A wins) = _____ P(Spinner B wins) = _____

2. Use your class totals to calculate the following experimental probabilities.

    P(Spinner A wins) = _____ P(Spinner B wins) = _____

3. a. Complete the matrix at the right for the *Spinner Game* in Activity 2.

    b. Are the outcomes in the matrix equally likely? Explain.

    A = Spinner A wins
    B = Spinner B wins

		Spinner B		
		3	5	7
Spinner A	2			B
	4			
	9	A		

4. Use the data in the matrix to calculate the following theoretical probabilities.

    P(Spinner A wins) = _____

    P(Spinner B wins) = _____

5. a. Compare the probabilities in Exercise 4 to those in Exercise 1.

    b. Compare the probabilities in Exercise 4 to those in Exercise 2.

6. Is the *Spinner Game* a fair game? Explain.

EXTENSION   Analyze the game in the Extension in Activity 2 using matrices.

Activity 4:   Paper-Scissors-Rock

PURPOSE   Introduce the use of a matrix to find experimental probabilities and a tree diagram to calculate theoretical probabilities. Reinforce fair games and use a matrix to find theoretical probabilities.

COMMON CORE SMP   SMP 2, SMP 3, SMP 4, SMP 5, SMP 7, SMP 8

GROUPING   Work in pairs.

GETTING STARTED   The *Paper-Scissors-Rock* game has been popular for many years. The two-player game is played as follows:

- Each player makes a fist.
- On the count of three, each player shows either *scissors* by showing two fingers, *paper* by showing four fingers, or *rock* by showing a fist.
- If scissors and paper are shown, the player showing scissors wins, since the scissors cut paper.
- If scissors and rock are shown, the player showing rock wins, since a rock breaks the scissors.
- If paper and rock are shown, the player showing paper wins, since paper wraps a rock.

PROBABILITIES USING A MATRIX

1.   Do you think *Paper-Scissors-Rock* is a fair game? Explain.

2.   Play the game 45 times. Each player should tally the outcomes in a *matrix* like the one below.

		Your Partner		
		Paper	**Scissors**	**Rock**
You	**Paper**			
	Scissors			
	Rock			

3.   Use the data in your matrix to calculate the following experimental probabilities.

P(you win) = _____ P(your partner wins) = _____ P(tie) = _____

4.   Use the probabilities in Exercise 3 to decide whether *Paper-Scissors-Rock* is a fair game. Explain your decision.

MORE EXPERIMENTAL PROBABILITIES

1. Use the data in your matrix to calculate the following experimental probabilities.

   a. P(you show rock) = _____

   b. P(your partner shows paper) = _____

   c. P(you show rock) $\times$ P(your partner shows paper) = _____

   d. P(you show rock and your partner shows paper) = _____

2. How do your answers to Exercise 1 Parts c and d compare?

3. Is P(you show scissors and your partner shows rock) about equal to P(you show scissors) $\times$ P(your partner shows rock)? Explain.

THEORETICAL PROBABILITIES USING A MATRIX

You can determine if the game is fair without conducting an experiment.

1. Complete the *matrix* at the right.

2. If the players choose the sign they show randomly, each of the nine outcomes in the matrix is *equally likely*. Find the following probabilities in this case.

   A = A wins
   B = B wins
   T = Tie

   P(A wins) = _____

   P(B wins) = _____

   P(Tie) = _____

		Player B		
		Paper	Scissors	Rock
Player A	Paper			A
	Scissors		T	
	Rock			

3. Based on the probabilities in Exercise 2, is *Paper-Scissors-Rock* a fair game? Explain.

4. Use the matrix to find the following theoretical probabilities.

   a. P(A shows rock) = _____   b. P(B shows paper) = _____

   c. P(A shows rock) $\times$ P(B shows paper) = _____

   d. P(A shows rock and B shows paper) = _____

5. How do your answers to Exercise 4 Parts c and d compare?

6. How does the theoretical probability in Exercise 4 Part d compare with the experimental probability you found in Exercise 1 Part d above?

PROBABILITIES USING A TREE DIAGRAM

Since *Paper-Scissors-Rock* can be thought of as a multi-stage experiment, it can be analyzed using a *tree diagram*.

1. Complete the tree diagram at the right.

2. What does the outcome PP in the tree diagram mean?

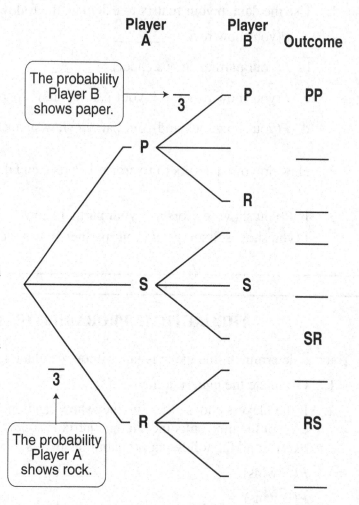

Consider the path leading to the outcome PP. Since the choices made by Player A and Player B are *independent* of one another, based on the probabilities along the path, we would expect the following:

- In $\frac{1}{3}$ of the games played, Player A will show paper.

- Player B will show paper in $\frac{1}{3}$ of the games in which Player A shows paper $\left(\frac{1}{3} \text{ of } \frac{1}{3}\right)$.

This shows that the probability of each outcome is the product of the probabilities along the path leading to the outcome.

Use the probabilities in the tree diagram to find the following theoretical probabilities.

3. $P(\text{PS}) = $ _____

4. $P(\text{SP or SR}) = $ _____

5. $P(\text{A wins}) = $ _____

6. $P(\text{B wins}) = $ _____

7. $P(\text{at least one player shows scissors}) = $ _____

Activity 5: How Many Arrangements?

PURPOSE	Introduce permutations and combinations.
COMMON CORE SMP	SMP 1, SMP 2, SMP 3, SMP 4, SMP 5, SMP 7, SMP 8
MATERIALS	Pouch: Colored squares
GROUPING	Work individually or in pairs.
GETTING STARTED	One square can be arranged in a row in one way.

Two different color squares can be arranged in a row in two distinct ways.

and

1. a. Use three different color squares. How many distinct ways can you arrange these three squares in a row? Record each arrangement as you make it. Enter the total number in the table.

Number of Squares	1	2	3	4
Number of Arrangements	1	2		

   b. How do you know you have found all the possible ways to arrange the squares?

   c. How is the number of arrangements of three squares related to the number of arrangements of two squares?

2. a. How many distinct ways do you think four different color squares can be arranged in a row? Why?

   b. Use four different color squares. Make as many distinct arrangements as you can with the squares. Record each arrangement as you make it.

   c. Enter the total number of arrangements in the table. How does this compare with your prediction in Part a? How is it related to the number of arrangements of three squares?

3. If you had *n* different color squares, how many distinct arrangements could you make? Justify your answer.

ARRANGEMENTS WITH LIKE SQUARES

1. a. Use two red squares and one green square. How many distinct ways can you arrange these three squares in a row? Record each arrangement as you make it.

   b. If all the squares in Part a had been different colors, how many arrangements would have been possible?

   c. The number of arrangements in Part a is what fraction of the number in Part b? How is this fraction related to the number of red squares?

   d. Explain why the number of arrangements of two red squares and one green square should be this fraction of the number of arrangements of three squares.

2. a. Repeat Exercise 1 Parts a–c using three red squares and one green square.

   b. If you started with four red squares and one green square, how many distinct ways do you think the five squares could be arranged in a row? Show how you made your prediction.

   c. Use four red squares and one green square. Make as many distinct arrangements as you can with the squares. How many arrangements did you find? How does this compare with the number you predicted in Part b?

3. How many distinct ways can you arrange each of the following sets of squares in a row?

   a. 2 red, 1 green, 1 blue   b. 2 red, 2 green   c. 3 red, 2 green

4. a. The number of arrangements of each set of squares in Exercise 3 is what fraction of the number of arrangements that would be possible if all of the squares had been different colors?

   b. How is each fraction in Part a related to the numbers of each color square in the set?

5. Suppose you have a set of squares some of which have the same colors. Explain how to find the number of distinct ways the squares can be arranged in a row.

In the preceding exercises, you have been working with permutations. A *permutation* is an arrangement of a group of items in which the order is important.

1. a. Suppose you have one red, one blue, one green, and one yellow square. Complete the tree diagram below to find the number of permutations that can be formed by choosing two of the squares and arranging them in a row.

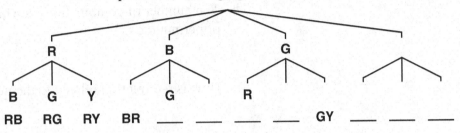

   b. How could you predict the number of permutations in Part a without using a tree diagram or listing them?

   c. How many permutations can be formed by choosing three of the squares? Explain how to find the number of permutations without listing the arrangements.

   d. Make a tree diagram to check your answer in Part c.

2. Now, suppose you want to find how many different pairs of squares can be chosen from a set containing one red, one blue, one green, and one yellow square.

   a. Is the pair consisting of the red square and the blue square (the pair RB) different from the pair BR? Explain.

   b. Does changing the order of the squares in a pair change the pair itself?

A selection of items in which order is not important is a *combination*.

3. a. For each arrangement in the tree diagram in Exercise 1 Part a on the previous page, cross out all the other arrangements that contain the same pair of squares. How many different combinations remain?

   b. The number of combinations is what fraction of the number of permutations?

   c. How is the fraction related to the number of squares in a pair?

4. Suppose you want to choose a group of three squares from a set containing one red, one blue, one green, and one yellow square.

   a. Use your tree diagram from Exercise 1d. Cross out the duplicate arrangements.

   b. How many different combinations are possible in this situation?

   c. What fraction of the number of permutations is this?

   d. How is the fraction related to the number of squares in a group?

5. How can you find the number of different possible subsets (combinations) of *m* items that can be selected from a set of *n* items?

Activity 6: Pascal's Probabilities

PURPOSE	Find theoretical probabilities in situations involving independent events using tree diagrams and Pascal's Triangle.
COMMON CORE SMP	SMP 2, SMP 3, SMP 4, SMP 5, SMP 7
GROUPING	Work individually or in pairs.

Suppose a family has three children. What is the probability that all three children are girls? Two are girls? Only one is a girl? None are girls? A tree diagram can be used to determine the theoretical probabilities of each number of girls.

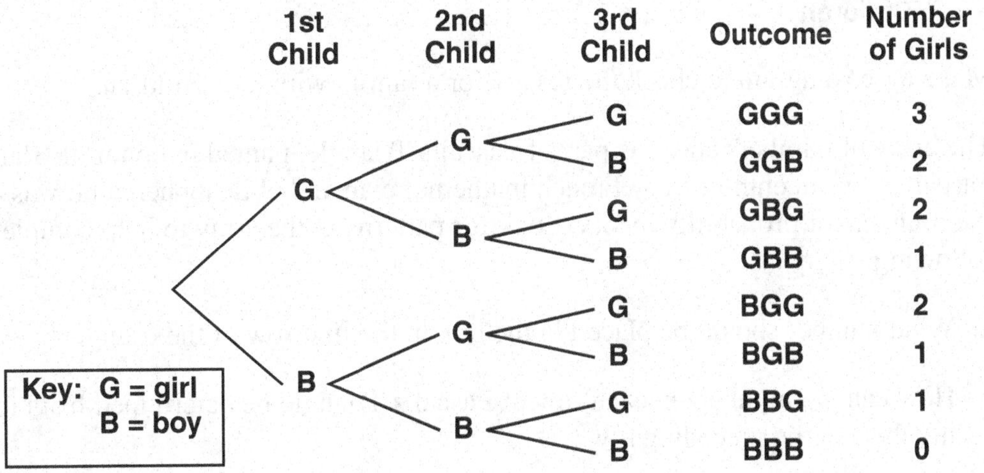

			Outcome	Number of Girls
1st Child	2nd Child	3rd Child		

Outcome	Number of Girls
GGG	3
GGB	2
GBG	2
GBB	1
BGG	2
BGB	1
BBG	1
BBB	0

Key: G = girl
     B = boy

1. Assuming that when a child is born the probability that it is a girl is 0.5, are the outcomes shown above equally likely? Explain.

2. Complete the following to summarize the results from the tree diagram.

Number of outcomes with   **Total outcomes**

0 girls   1 girl   2 girls   3 girls
↓   ↓   ↓   ↓

_____ + _____ + _____ + _____   =   _____

3. Use the results above to determine the following theoretical probabilities for the 3-child family.

$P(0 \text{ girls}) =$ _____ $P(1 \text{ girl}) =$ _____

$P(2 \text{ girls}) =$ _____ $P(3 \text{ girls}) =$ _____

Tree diagrams can also be used to find the outcomes for families with other numbers of children, but such diagrams quickly become very large and complex. The results for families with 1, 2, 3, 4, and 5 children are shown in the diagram below. The results are listed in order from the number of outcomes with 0 girls to the number with all girls.

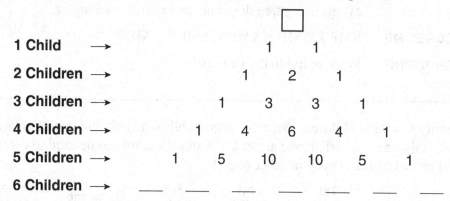

1. Make a tree diagram to check the results for a family with four children.

2. The array of numbers above is part of Pascal's Triangle, named in honor of Blaise Pascal, a seventeenth-century French mathematician and philosopher, who was one of the founders of probability theory. Look for patterns in the array to help complete the following.

   a. What number should be placed in the box in the first row of the triangle?

   b. How can the numbers in each row of Pascal's Triangle be determined from the numbers in the preceding row?

   c. Complete the seventh row of Pascal's Triangle.

3. Use the numbers in the sixth row of Pascal's Triangle to calculate the theoretical probability of each number of girls in a family with five children.

4. Use the numbers in the seventh row to help answer the following questions. What is the probability that in a family with six children

   a. there are exactly three girls?   b. there are at least three girls?

   c. there is exactly one boy?   d. there are more than two boys?

5. Use Pascal's Triangle to calculate the theoretical probability for each number of girls in a family with nine children. Explain your procedure.

Activity 7: Simulate It

PURPOSE	Analyze probability situations using simulations.
COMMON CORE SMP	SMP 1, SMP 2, SMP 3, SMP 4, SMP 5, SMP 7, SMP 8
MATERIALS	Other: One die
GROUPING	Work individually or in pairs.

DESIGNING A SIMULATION

Robin Hood and Maid Marian are having an archery contest. They alternate turns shooting at a target. The first person to hit it wins. The probability that Robin hits the target is $\frac{1}{3}$ and the probability that Marian hits it is $\frac{2}{5}$. Since Marian is the better archer, Robin shoots first. What is the probability that Marian will win the contest? On average, how many shots will be necessary to determine a winner? To find the answers, you can simulate the problem.

Step 1: Select a model.

*   To simulate Robin's shot, roll a die. If the result is a 1 or 2, he hits the target. Otherwise he misses.

*   For Marian's shot, roll a die. If the outcome is a 1 or 2, she hits the target. If it is a 3, 4, or 5, she misses. If it is a 6, roll the die again.

Step 2: Conduct a trial and record the result.

A trial consists of rolling a die to alternately simulate a shot for Robin and then one for Marian until someone hits the target.

Example:	Shot	Archer	Die Roll	Result
	1	Robin	4	Miss
	2	Marian	6	Roll Again
			5	Miss
	3	Robin	1	Hit—Robin Wins!

Step 3: Repeat Step 2 until the desired number of trials is completed.

Complete 30 trials and record the results in a table like the one below.

Trial	Winner	Number of Shots	Trial	Winner	Number of Shots
Example	Robin	3			

Step 4: Interpret the results.

Based on the results of the 30 trials, what is the probability that Robin wins the contest? What is the average number of shots for the 30 trials?

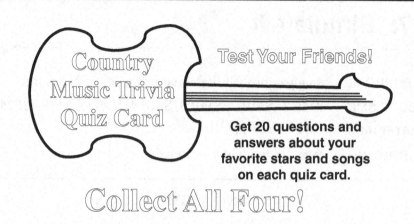

Get 20 questions and answers about your favorite stars and songs on each quiz card.

Collect All Four!

As part of a promotional campaign, a cereal manufacturer packages one Country Music Trivia Quiz Card inside each specially labeled box of cereal.

A country music aficionado wants to collect a complete set of the Country Music Trivia Quiz Cards. He wants to estimate how many boxes of cereal he will need to purchase in order to collect all four Quiz Cards.

1. Describe how the card obtained by purchasing one box of cereal can be modeled by rolling a die.

2. What would make up a trial?

3. Complete 20 trials and record the results in a table like the one below.

Trial	Number of Boxes	Trial	Number of Boxes

4. On average, how many boxes of cereal will the aficionado have to purchase in order to collect a complete set of four Country Music Trivia Quiz Cards?

5. Describe a model that could be used if one card was twice as likely to be in a box of cereal as the others.

6. In this case, what would make up a trial?

EXTENSION   Suppose Robin and Marian change the rules of their tournament so that the winner is the person who wins two out of three contests. Draw a tree diagram to show the possible outcomes of the tournament. Use the results from the archery simulation to assign probabilities to the branches and determine the probability that Marian wins the tournament.

Chapter Summary

Probabilities are used to help interpret and understand situations involving uncertainty. There are two kinds of probabilities: experimental and theoretical. *Experimental probabilities* are determined by observing the outcome of a situation or experiment over a large number of trials. For example, meteorologists collect weather data over an extended period of time. When they say that there is a 20% chance, or probability, that it will snow on a given day, they mean that, in the past, it snowed on 20% of the days with similar weather conditions.

Theoretical probabilities are based on mathematical analysis of the possible outcomes of an experiment rather than on observation of the outcomes. A fundamental concept of probability is that over a large number of trials, the experimental probability of an event will approach the theoretical probability. This idea was emphasized by having you compare your individual results from an experiment with the theoretical results and then combining your results with other people's to obtain the results for a larger number of trials and repeating the comparison. In general, as the number of trials increases, the experimental probability will more closely approximate the theoretical probability.

The concept of probability was introduced in Activity 1 by building on intuitive ideas about chance. Activity 2 introduced the idea of a *fair game*. The intuitive notion is that a game is fair if the chances of winning are equal to the chances of losing, that is, if all players in a game have the same chance of winning. The notion of a fair game was explored further in Activities 3 and 4 where a game was analyzed using matrices and probability trees, tools that are useful in determining probabilities.

Many of the basic concepts of probability were introduced in Activities 2 and 3. You learned that a probability is a ratio that expresses the likelihood of something happening. You also explored the distinction between experimental and theoretical probabilities. Some of the basic ideas introduced in these activities are that

$$\text{Experimental probability} = \frac{\text{number of times an event occurs}}{\text{total number of trials}}$$

and that if the outcomes of an experiment are equally likely, then

$$\text{Theoretical probability} = \frac{\text{number of outcomes making up the event}}{\text{total number of possible outcomes}}.$$

If an event **cannot happen**, its probability is 0, and if an event **is certain to happen**, its probability is 1. If A and B are *complementary events*, then $P(A) = 1 - P(B)$. And finally, if C and D are *mutually exclusive events*, that is, events that cannot occur simultaneously, then $P(C \text{ or } D) = P(C) + P(D)$.

In Activity 4, a tree diagram was used to determine the probabilities of the outcomes of a multi-stage experiment. In the activity, you learned that the probability of each outcome is the product of the probabilities along the path leading to it.

The use of permutations and combinations to count the outcomes of an experiment was introduced in Activity 5, and Activity 6 showed how the probabilities of certain events could be determined using Pascal's Triangle.

Many real-world probability problems are too complex to analyze theoretically and too difficult, time consuming, or expensive to observe actual trials. In these cases, the solution can often be found by simulating the problem. Activity 7 developed the techniques used to design and conduct a *simulation*. Because a large number of trials must be conducted to obtain accurate results by simulating an experiment, computers are often used to carry out the simulations.

COLOR MANIPULATIVES

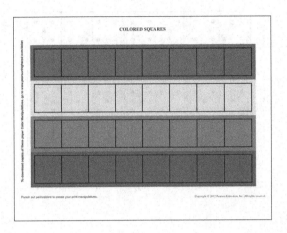

COLORED SQUARES

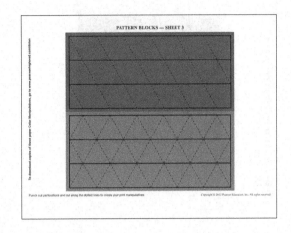

PATTERN BLOCKS — SHEET 3

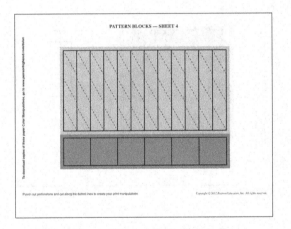

PATTERN BLOCKS — SHEET 4

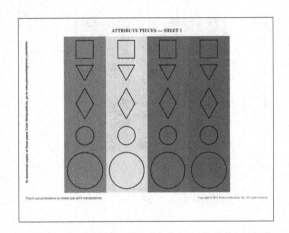

ATTRIBUTE PIECES — SHEET 1

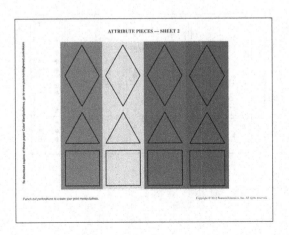

ATTRIBUTE PIECES — SHEET 2

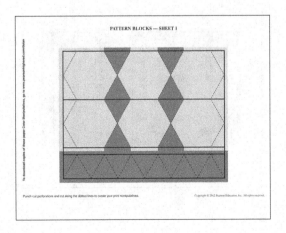

PATTERN BLOCKS — SHEET 1

To download full size paper Color Manipulatives, go to www.pearsonhighered.com/dolan

COLOR MANIPULATIVES

PATTERN BLOCKS — SHEET 2

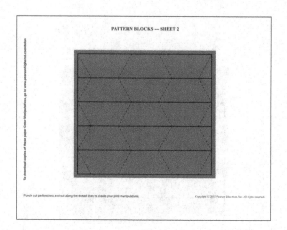

DOUBLE-SIX DOMINOES — SHEET 2

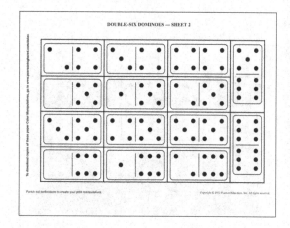

CUISENAIRE® RODS

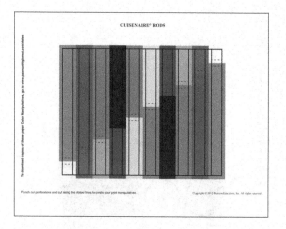

BASE TEN BLOCKS (A)

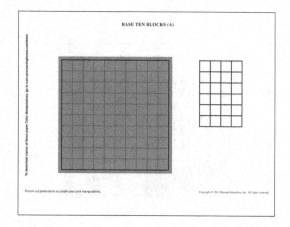

BASE TEN BLOCKS (B)

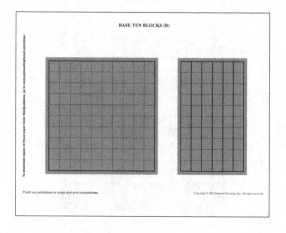

DOUBLE-SIX DOMINOES — SHEET 1

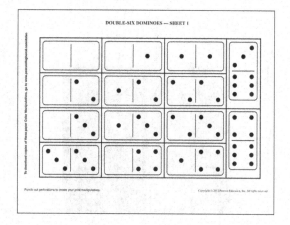

ACTIVITY MASTERS

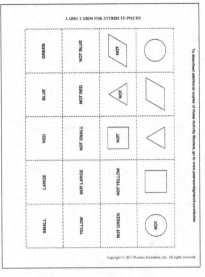

LABEL CARDS FOR ATTRIBUTE PIECES

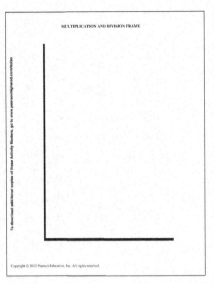

MULTIPLICATION AND DIVISION FRAME

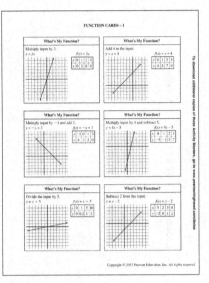

FUNCTION CARDS – 1

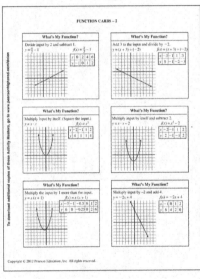

FUNCTION CARDS – 2

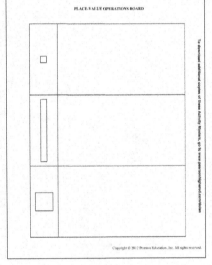

PLACE-VALUE OPERATIONS BOARD

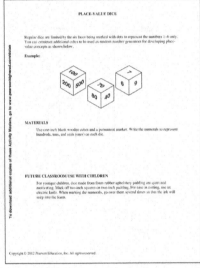

PLACE-VALUE DICE

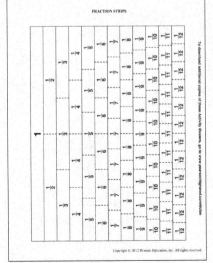

FRACTION STRIPS

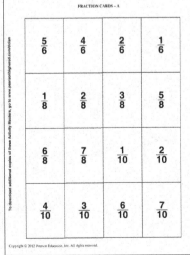

FRACTION CARDS – A

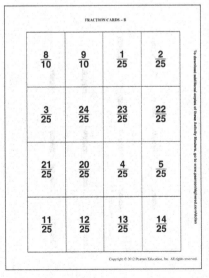

FRACTION CARDS – B

ACTIVITY MASTERS

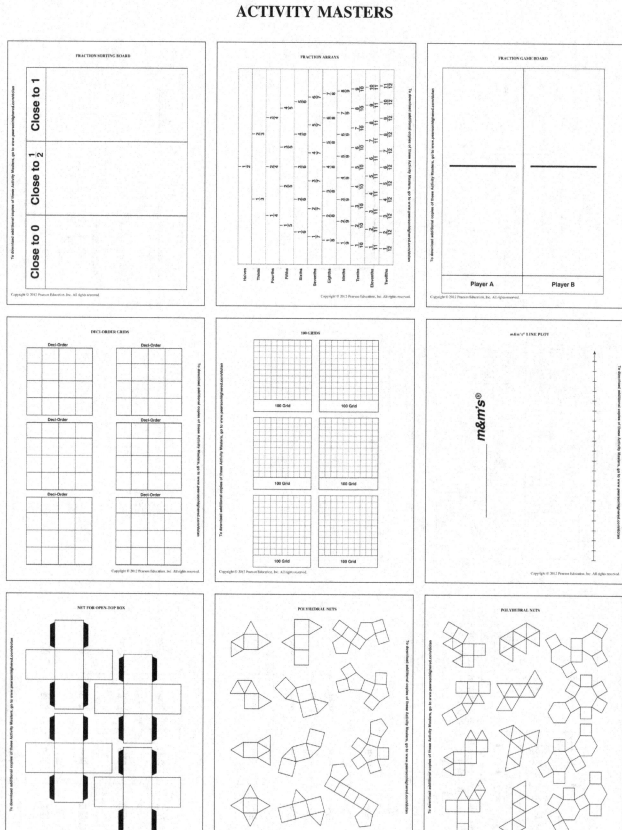

ACTIVITY MASTERS

GEOBOARDS DOT PAPER
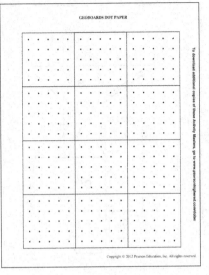

PYTHAGOREAN PUZZLES
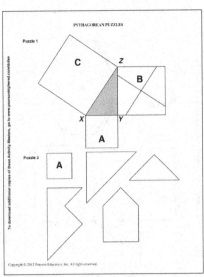

HALF CENTIMETER GRAPH PAPER
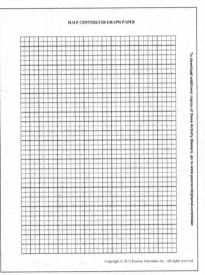

CENTIMETER GRAPH PAPER
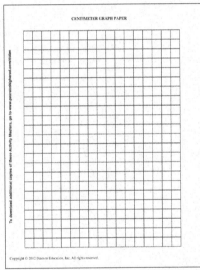

DOT PAPER
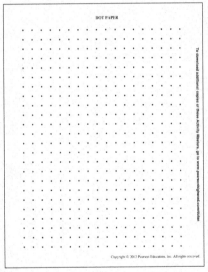

REGULAR POLYGONS
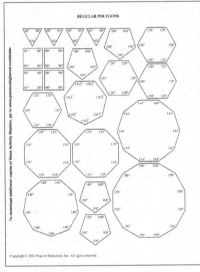

NET FOR PRISM
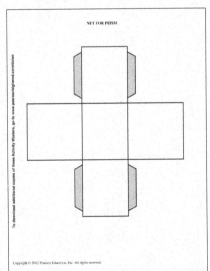

NET FOR PYRAMID
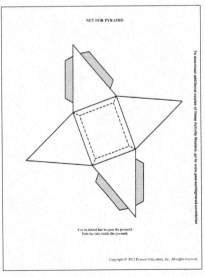

Cut on dotted line to open the pyramid.
Fold the tabs inside the pyramid.

NET FOR CYLINDER

Tab for overlap to tape

ACTIVITY MASTERS

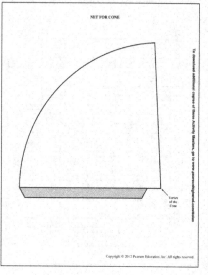

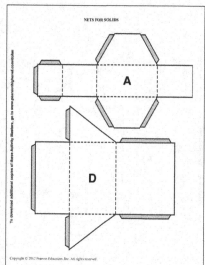

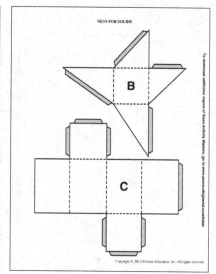

Index

Credits

Chapter 1, p. 01   "George Polya, How to Solve It, 1957"

Chapter 1, p. 01   Reprinted with permission from Principles and Standards for School Mathematics. Copyright (c) by the National Council of Teachers of Mathematics. All rights reserved.

Chapter 1, p. 01   Reprinted with permission from Principles and Standards for School Mathematics. Copyright (c) by the National Council of Teachers of Mathematics. All rights reserved.

Chapter 2, p. 23   Reprinted with permission from Principles and Standards for School Mathematics. Copyright (c) by the National Council of Teachers of Mathematics. All rights reserved.

Chapter 2, p. 23   Reprinted with permission from Curriculum Focal Points for Prekindergarten through Grade 8 Mathematics. Copyright (c) by the National Council of Teachers of Mathematics. All rights reserved.

Chapter 3, p. 37   Reprinted with permission from Principles and Standards for School Mathematics. Copyright (c) by the National Council of Teachers of Mathematics. All rights reserved.

Chapter 3, p. 55   "Reprinted with permission from Developing Number Sense, Addenda Series, Grades 5, p. 8. Copyright (c) by the National Council of Teachers of Mathematics. All rights reserved. "

Chapter 4, p. 57   Reprinted with permission from Principles and Standards for School Mathematics. Copyright (c) by the National Council of Teachers of Mathematics. All rights reserved.

Chapter 4, p. 57   Reprinted with permission from Curriculum and Evaluations Standards for School Mathematics. Copyright (c) by the National Council of Teachers of Mathematics. All rights reserved.

Chapter 5, p. 75   Reprinted with permission from Principles and Standards for School Mathematics. Copyright (c) by the National Council of Teachers of Mathematics. All rights reserved.

Chapter 5, p. 75   Reprinted with permission from Curriculum Focal Points for Prekindergarten through Grade 8 Mathematics. Copyright (c) by the National Council of Teachers of Mathematics. All rights reserved.

Chapter 6, p. 89   Reprinted with permission from Principles and Standards for School Mathematics. Copyright (c) by the National Council of Teachers of Mathematics. All rights reserved.

Chapter 6, p. 89   © Copyright 2010. National Governors Association Center for Best Practices and Council of Chief State School Officers. All rights reserved.

Chapter 7, p. 111   Reprinted with permission from Curriculum Focal Points for Prekindergarten through Grade 8 Mathematics. Copyright (c) by the National Council of Teachers of Mathematics. All rights reserved.

Chapter 7, p. 111   Reprinted with permission from Principles and Standards for School Mathematics. Copyright (c) by the National Council of Teachers of Mathematics. All rights reserved.

Chapter 7, p. 111   Reprinted with permission from Principles and Standards for School Mathematics. Copyright (c) by the National Council of Teachers of Mathematics. All rights reserved.

Chapter 8, p. 131   Reprinted with permission from Navigating through Algebra in Grades 3, p. 5. Copyright (c) by the National Council of Teachers of Mathematics. All rights reserved.

Chapter 8, p. 131   Reprinted with permission from Principles and Standards for School Mathematics. Copyright (c) by the National Council of Teachers of Mathematics. All rights reserved.

Chapter 9, p. 151   Reprinted with permission from Principles and Standards for School Mathematics. Copyright (c) by the National Council of Teachers of Mathematics. All rights reserved.

Chapter 9, p. 151   Reprinted with permission from Principles and Standards for School Mathematics. Copyright (c) by the National Council of Teachers of Mathematics. All rights reserved.

Chapter 9, p. 151   Reprinted with permission from Principles and Standards for School Mathematics. Copyright (c) by the National Council of Teachers of Mathematics. All rights reserved."

Chapter 9, p. 152   "A Mathematician's Delight, W. W. Sawyer"

Chapter 10, p. 175   Reprinted with permission from Principles and Standards for School Mathematics. Copyright (c) by the National Council of Teachers of Mathematics. All rights reserved.

Chapter 10, p. 175   © Copyright 2010. National Governors Association Center for Best Practices and Council of Chief State School Officers. All rights reserved.

Chapter 11, p. 193   Reprinted with permission from Principles and Standards for School Mathematics. Copyright (c) by the National Council of Teachers of Mathematics. All rights reserved.

Chapter 12, p. 215   Reprinted with permission from Principles and Standards for School Mathematics. Copyright (c) by the National Council of Teachers of Mathematics. All rights reserved.

Chapter 12, p. 215   Reprinted with permission from Curriculum Focal Points for Prekindergarten through Grade 8 Mathematics. Copyright (c) by the National Council of Teachers of Mathematics. All rights reserved.

Chapter 13, p. 231   Reprinted with permission from Principles and Standards for School Mathematics. Copyright (c) by the National Council of Teachers of Mathematics. All rights reserved.

Chapter 13, p. 237   Data from The World Almanac and Book of Facts 2011

Chapter 13, p. 247   Data from National Climatic Data Center

Chapter 13, p. 252   Data from NEA Rankings and Estimates Update—A Report of School Statistics—(Fall 2004)

Chapter 14, p. 257   Reprinted with permission from Principles and Standards for School Mathematics. Copyright (c) by the National Council of Teachers of Mathematics. All rights reserved.

Chapter 14, p. 257   © Copyright 2010. National Governors Association Center for Best Practices and Council of Chief State School Officers. All rights reserved.